CHICKEN
BREEDS AND CARE

Expert practical guidance on keeping chickens
plus profiles of all the major breeds.

Frances Bassom

Reference Only

INTERPET
PUBLISHING

Reprinted 2009, 2010, 2011,
2012 (twice), 2013

ISBN: 978 1 84286 212 4

Credits
Created and compiled: Ideas into
Print, Claydon, Suffolk
IP6 0AB, England.
Design and Prepress: Stuart
Watkinson, Ayelands, Longfield,
Kent DA3 8JW, England.
Computer graphics:
Stuart Watkinson
Photography: Geoff Rogers
© Interpet Publishing
(Also see Credits page 208.)
Production management:
Consortium, Poslingford, Suffolk
CO10 8RA, England.
Print production: 1010 Printing
International Ltd

Printed and bound in China

Contents

Part One: Practical Section 8-81

Part Two: Chicken Breed Profiles 82-199

Author

Frances Bassom has had a fascination with all things poultry related from a very early age. Eventually, an interest in rare and exhibition breeds led to the purchase of a trio of Sebrights, rapidly followed by many other breeds, with an emphasis on those classed as rare, such as Lakenvelders and Vorwerks. In 1981, Frances became one of the earliest members of the Surrey Poultry Society, of which she is currently chairman. She is a member of the Rare Poultry Society and prefers to describe herself as a breeder of poultry rather than an exhibitor. She has developed a new variant of the Vorwerk Bantam and is currently endeavouring to preserve and promote a chocolate gene found in the Minorca Bantam.

Contents

"There is a style in Chickens as well as in anything else. A new breed will always have a great many admirers at first and great claims will be made for its superior qualities. The poultry men who have stock and eggs to sell will secure high prices for their output. Very soon however, the real value of a new breed will be known and it will be on the same basis as the older breeds." Claude Harris Miller, 1911.

Poultry keeping is booming

Poultry are popular. Hardly a week seems to go by without a magazine or newspaper carrying an article about another celebrity's new-found love of keeping poultry. Not since the heady days of the 1850s, when 'hen fever' was at it height and people vied with each other to purchase new breeds of poultry for enormous sums of money, have our feathery friends been so in demand.

For those people who had been quietly keeping poultry for many years, it initially came as something of a shock to discover more and more people deciding that they wished to add poultry keeping into their lives. After all, prior to all this recent interest, they had been classed as the odd ones out!

One of the reasons for this rise in popularity stems from the fact that people have realised that they need to become more curious as to how their food is produced. Added to this is the growing awareness of the large distances most food has travelled to reach our plates. It is estimated that 23% of our carbon footprint is taken up in food miles. Eating eggs, meat and vegetables from your garden or allotment, not only reduces your carbon footprint, but also adds to a healthy way of life and provides an enjoyable pastime the whole family can get involved in.

The rescue of battery hens from cages, allowing them to live out the rest of their productive lives cage-free, is perhaps one of the fastest-growing ways that people are being introduced to the pleasure of having eggs from your own hens. With their missing feathers and trimmed beaks they look a far cry from the fancy show birds. Yet once they have regrown their feathers and are living happily in your garden, you will discover that as well as laying eggs, they also arrive fully equipped with a personality. Caring for such hens is a wonderful introduction to domestic poultry keeping.

Although at first you may become involved in poultry through keeping a few laying hens, sooner or later the chances are that you will visit somewhere that has other, more exotic-looking breeds of poultry and be tempted to enlarge your flock. Not all breeds are suitable for novices. If you already keep laying hens you can hardly be classed as a novice, but some breeds may have a temperament trait that you are unaware of; or maybe their care will consume more time than you can afford if you have a busy lifestyle.

This book provides not only the basic information you need to care for your birds, but also introduces you to a whole range of breeds that you may not have known existed. From the ultra-rare to the fluffy and cute; from the solid, reliable utility types to the fine-boned thoroughbreds of the poultry fancy, somewhere out there is a breed of chicken that will fit perfectly into your back garden — a bird that will meet all your requirements and spark enough interest to convince you that keeping poultry is definitely the hobby for you.

From the largest Asiatics down to the diminutive True Bantams, there is a breed of poultry for every size of garden. Here, a Dark Brahma male poses with a Dutch pair, emphasising the size difference between the two.

History of domestic breeds

To start this review simply by repeating what has been written about the origins of domestic poultry in numerous books published over the last 100 or so years would be to ignore recent developments in the world of science.

For many years it was assumed that all poultry derived solely from the Red Jungle Fowl *(Gallus bankiva)*, which is found over quite a large area, ranging eastwards from northeast India and southern China into many Southeast Asian countries.

However, this assumption of a one-bird origin has always had its detractors. Many long and heated discussions have taken place over the years, with some people firmly holding to the belief that there had to be more than one ancestor. This conviction was based on some of the more unusual attributes found in certain breeds.

As a result of recent developments in DNA analysis, scientists are starting to unveil evidence to support those doubters. As recently as 2008, scientists reported that the yellow skin gene found in domestic poultry does not originate from the Red Jungle Fowl. A more likely origin for this gene is from the Grey Jungle Fowl *(Gallus sonnerati)*. This suggests a possible cross between the two lines thousands of years ago.

Unfortunately, the origin of domestic poultry, although fascinating to some people, has never rated very highly on the archaeological radar of importance. The result is that finds of chicken bones have usually been ignored as of little or no interest. Since the nineteenth century, finds of bones attributed to various species of *Gallus* have been turning up in Neolithic sites all over Europe and are only now attracting more interest. This may help shed light on the matter or may confuse the picture even more; only time will tell if these discoveries will add fuel to future debates. However, we can be sure that the complete story of the origins of the chicken are set to run for a number of years yet.

Left: *Pure Red Jungle Fowl are never as tame as any forms of domestic poultry and need to be kept in very large, aviary-like emclosures.*

Below: *In England, the ancient Celts were producing zoomorphic brooches in the form of cockerels from around 150 A.D.*

Early beginnings

Bones of domestic fowl dating to around 5000 B.C. have been recovered during archaeological excavations at Cishan in northern China. Possibly of more relevance to modern domestic breeds of fowl is the fact that other chicken bones, later identified as being not only of a Red Jungle Fowl type, but also a Malay-type fowl and another of intermediate size (possibly of similar build to an Asil), were found in northeast Thailand and dated around 3500 B.C.

For documented evidence of poultry domestication we only have to look to the ancient writings of Pliny, Aristotle, Siculus and Columella. A wallpainting found in Egypt in the tombs of Petosiris in the fourth century B.C. depicts a cockerel being carried in a procession. Julius Caesar mentioned fowl being kept when he invaded Britain in 55 B.C. Therefore, we can quite safely say that poultry keeping has had its followers since the very early histories of civilisation. The most recent suggestions are that domestic fowl were well established in China by around 6000 B.C.

Spreading by trade and conflict

We can only speculate as to how they spread across the world, but we can assume that poultry

were taken as a portable food source into wars and on long sea voyages. Of course, cockerels would also have been taken as a way to pass the time, cockfighting being a regular diversion in ancient times, and they would have provided a readily saleable item of trade.

Only in the latter part of the eighteenth century do we start to get really detailed and accurate recording of breeds, their different physical characteristics and utility attributes. Yet by delving through ancient writings we can find clues that some breeds, readily identifiable from their descriptions, have remained very true to type for countless generations.

Above: *The Dunghill Cock was a generic name given to the barnyard fowl of the seventeenth century. Interestingly, this one shows the triple comb most often seen in Asiatic breeds.*

9

Which type of chicken is best for you?

Without doubt, the best hen that anyone can own is a healthy one, so before you buy your birds, familiarise yourself with the points to look out for when choosing poultry (see page 12).

Many different breeds of chicken have been developed as the result of careful selection over hundreds of years. Studying the profiles of the breeds featured in this book should help you to choose the best bird to suit your own personal situation and tastes. They are discussed in the following categories:

• **Light breeds** are usually active, alert and intelligent; in addition they will usually produce a good number of eggs.

• **Heavy breeds** are physically larger, and usually more docile. In addition, they are often utility breeds producing both eggs and meat.

• **Bantams** are breeds that either have no Large Fowl counterpart or are miniatures of a large breed. Some of them are amongst the most attractive and charming poultry breeds available. Bantams are an ideal choice where space is limited.

• **Game breeds** are the gladiators of the poultry world, Strong and powerful they now battle for prizes in the show ring rather than in the cock-pit.

If you want birds for both eggs and meat, then select one of the many utility strain breeds that will perform both roles, such as Welsummer, Marans or Sussex.

There is little point is acquiring a breed solely on its show, egg laying or utility

The Sussex is a very popular Utility Heavy Breed.

The Vorwerk is an active foraging Light Breed.

The Pekin is possibly the best-loved of the True Bantams.

merits if you don't really like the look or character of the breed. Time taken in researching the right breed for you will pay dividends in the long run and enable you to get the most enjoyment out of your hobby.

Hens for showing and breeding

If you want to show and breed from your birds, you can select any of the pure breeds.

Trends in poultry keeping fluctuate, and a breed that is popular one year can be teetering on the verge of extinction the next. Even a small back garden setup can provide a vital oasis for some of the rarer pure breeds.

Hybrid hens

If keeping poultry solely for eggs there is no better bird than one of the many varieties of commercial hybrids. Hybrid hens

The Modern Game is an excellent exhibition Game Breed.

are no longer just the standard brown hen that has decorated poultry yards since the 1950s. They are now available in a range of pretty colours, and they also lay different-coloured eggs, ranging from brown to blue.

Remember that hens will lay eggs just as well without a cockerel present to distract them. You only need to keep a cockerel if you want to breed from your birds.

What age chicken should I buy?

If egg production is your main objective, the optimum age to acquire poultry is when the bird is approaching the point when it will start to lay. This is called 'point of lay', usually abbreviated in advertisements to POL. Hybrid birds will be 16-20 weeks of age, but Heavy pure breeds can be older.

If you are looking to acquire birds for breeding, then hens over one year old and just past their first moult are a better choice, as they have proved themselves to be strong stock, able to meet the rigours of both a laying season and a moult.

Hybrid chickens are always the best breeds for egg laying.

11

Where do I buy poultry?

All types of poultry are advertised in the poultry-related magazines on sale monthly from newsagents, and most of these publications contain a breeder's directory. The notice board in your local agricultural merchant will also often yield good contacts.

Possibly the most popular source of commercial hybrids are rescued ex-battery hens. Although these birds do need a bit of extra care at first, they soon regrow their feathers and settle down to a happy cage-free life.

When looking for pure breeds you can do no better than visit a local poultry or agricultural show, where you can study the birds first hand and talk to the breeders. It is also worth making contact with a poultry society; there is sure to be one located somewhere near you. The internet is a great source

Crests add an
ornamental feature.

Beards and
muffling
need to stay
clean.

Healthy chicken checklist

- *Look for a bright, active-looking bird, with a good posture – not drooping or hunched.*
- *Bright bold eyes, with the colour of the iris matching in each eye. Check that the pupils are even in shape and not blending into the iris. The eyes should not look wet or bubbly.*
- *The nostrils should show no signs of mucus. Hold the bird and gently press the side of each nostril to check for this. If you are unsure, place your head to the bird's back and listen; there should be no hint of wheezing.*
- *Check that the sinuses just above the nostrils are not swollen or puffy.*
- *Ears should not have any cheeselike substance in them, as this is a sign of infection.*
- *The comb should be a bright colour and not shrunken or shrivelled. A comb that is purple instead of red can indicate circulation problems and general ill health. But do be aware that some breeds have naturally darker-coloured combs.*
- *The legs should be smooth, with no raised scales or rough, crusty yellow patches.*
- *The feathers around the vent area should be clean. Dirt here can indicate parasites, and a closer look may reveal mites or lice.*
- *Check the crest and beard in birds that have them to ensure that there is no evidence of northern fowl mite lurking there.*
- *Handle the bird; it should feel firm for its size, with a reasonable covering of flesh on each side of the breastbone. Some allowances have to be made for birds of Light Breeds that will be less covered in the breast area, but the breastbone should never feel exceedingly prominent, with no flesh at the sides at all.*

of information, and doing a search on the words 'poultry club' or 'poultry society' and your home town will be sure to find a contact near you.

Although livestock markets often have a section for poultry, stock from these sources can introduce all sorts of problems unless you know what you are looking for. Novice poultry keepers should avoid them, as the temptation to buy a sweet, fluffy – but possibly unhealthy – bird can often overrule commonsense.

Know your chicken

Nostril

Comb

Secondary flight feathers

Tail sickle

Ear

Saddle hackle

Beak

Earlobe

Neck hackle

Wattles

Breast

Wing covert

Keel

Primary flight feathers

Side hangers

Scales

Hock

Vent

Toenail

Shank

Abdomen

Spur

Housing chickens

If you look at the many Victorian poultry books still in existence, one thing immediately stands out: the design of poultry houses is very much the same today as it was over 100 years ago. This is because poultry have certain requirements that must be met in order for them to live contentedly in confinement.

There is a huge array of poultry houses on the market and the many countryside-related magazines are full of advertisements for houses of all kinds. A good poultry house should be a suitable size and shape for the number of birds you want to put in it. It should be vermin- and weatherproof, but with adequate ventilation. It will need a small door called a pop hole to allow the birds access to and from the outside. It will also need nest boxes and perches. Importantly, you should be able to access it easily so that you can clean it out and catch any birds. This will usually mean a door wide enough to

Above: *Houses are frequently built on legs or set on wooden skids to discourage rats and mice from taking up residence underneath the house. It also protects the floor from possible rot caused by resting on damp ground.*

allow you to reach into the corners.

Some houses are supplied with runs attached, others are designed so that you either keep the birds on free range or build a permanent run. The one you choose will depend on your own situation.

The size of the house

Housing manufacturers often say that a house is designed for a certain number of birds, stating, for example, 'suitable for four to six birds'. Bear in mind that they are usually thinking of commercial layers, which are a small-bodied, leghorn-type bird. If you plan to keep either larger or much smaller breeds,

Above: *A small house like this is ideal for a broody and chicks. The run roof can be easily removed.*

you must take this into consideration. You should allow about 1m² of floor space for every five birds.

Another thing to note is that the quoted size will usually be for birds kept on free range. If your birds are to be confined to a run, they will spend more time in the house than if they were able to shelter from inclement weather under trees and hedges. As a general rule, you should stock to the lower number of the figures quoted.

The shape of the house

A square or rectangular house, with a large side- or rear-opening door, is a very good option as a first chicken house. Houses designed on the triangular shape are easy to build and save on materials. (Originally, they were built this shape to stop sheep jumping on them, which is something you may not have to worry about.) Although traditional, this

Above: A house such as this meets the requirements of keeping birds under cover in the event of an avian influenza outbreak.

This house is very suitable for a trio of Bantams. Keep your poultry house in good condition by treating it with a safe wood preservative.

Left: Having the sleeping quarters of the house raised from the ground allows easy access for cleaning if you have a bad back.

15

Above: *Bituminised corrugated roofing such as this is now widely used as it does not tend to harbour parasites.*

Left: *Attached runs should always have their own access, here via removable mesh panels.*

shape has drawbacks and does not make the most ideal accommodation. In triangular houses, often called arks, there is always a percentage of unusable space, as the pitched roof becomes very low in the eaves. However, these coops are useful for a broody and chicks, as the chicks are small enough to get into these spaces. The problem of the low eaves can be overcome by incorporating an additional side wall, and these modified arks are a much better choice than the standard shape.

Houses that open only via a hinged roof can prove difficult to clean out. If you are considering one of these, open the roof and look inside to make sure that you will be able to remove the litter and droppings easily. Thoroughly cleaning a house only via the roof can be a strain on your back.

Materials

Whatever type of house you choose it should be solid enough to keep out foxes and other predators. Wood is strong and also a very good insulator, and has traditionally been used for poultry housing. The wood should be treated against rot with a suitable timber preserver. Whatever the type of wood, it should be sufficiently thick to last a number of years. Unfortunately, wood does harbour parasites unless treated on a regular basis.

If you choose a plywood house, the plywood should be made with a water- and boil-proof glue, as this will not warp as readily as other cheaper versions. This is usually indicated by the letters WBP on the description of the wood.

Plastic houses

Recently, designs of poultry houses made of plastic have come on the market. If you choose one of these, do make sure that it incorporates a twin wall to help keep it cool in summer and warm in winter. Most of these houses are designed for no more than a couple of laying hens, as their size is unsuitable for more. Their modern design and bright colours have proved popular in urban environments, where many families just require a couple of hens for egg laying. They are easy to clean and tend not to harbour pests. They also retain a high secondhand value if you ever decide to sell the house.

Ventilation

A light, airy house is a much pleasanter and healthier environment for hens than a dark and stuffy one. Keepers realised this many years ago; indeed the Victorians were recommending light and air for their poultry long before they realised that these conditions were important in hospitals for humans!

Ventilation in a hen house is most important, but novices easily overlook this requirement. Without adequate ventilation, a house will quickly become stuffy, unhygienic and damp, forming an ideal breeding ground for respiratory problems amongst the flock.

Ventilation can be in the form of a window covered in wire mesh, or via specially designed grills and air holes. Alternatively, you can incorporate a corrugated-type roof in which the many ridges provide gaps. Never provide ventilation by way of stray draughts from badly fitting walls and doors. Locate the ventilation at the top of the house to permit the stale warm air to escape.

If the house is fitted with a mesh window with a shutter, do not close the shutter just because the weather is cold. Feathers give poultry a good tolerance to the cold, providing they are dry, but the animals will suffer if their environment is too humid. It is better to have an adequate overhang of roof on the hen house to keep out inclement weather rather than totally close the shutter.

In summer the sun rises very early and will wake your birds. Unless you are prepared to get up and let them out as soon as dawn breaks, or have fitted

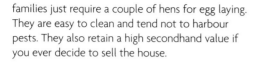

Above and right: These modern plastic houses are popular with newcomers to keeping chickens. They are available in a range of bright colours and are easy to keep clean. They have a built-in nest box, as well as perches. Newly laid eggs are easy to retrieve from the nest box. This model is for two laying hens.

your house with an electronic opening device, your hens will be spending part of their day waiting for your arrival. A window will provide both light and air for them, as in the summer months temperatures can rise rapidly. Furthermore, the added daylight provided by a window will encourage them to lay.

The floor

The floor can be either slatted or solid. A slatted floor will need some form of predator protection around the house itself. Many a poultry keeper has discovered to their horror that foxes can get under slatted houses and bite at the feet of their poultry, with fatal results. A slatted floor has the advantage that any droppings and mess fall through the slats and keep the floor clean. When the house is moved the droppings can be shovelled up from the ground and put on the compost heap.

A solid wooden floor is far warmer for the hens in winter, especially with a good bed of shavings. Given a choice, you will find that hens much prefer standing on a solid floor to a slatted one. It is possible to combine the best of both systems by fitting a temporary solid floor over the slats in the coldest weather, and removing it at other times.

The roof

The roof should be made of a weatherproof material and have a good overhang to prevent driving rain or snow accessing the house. Most traditional houses use mineral felt over wood, but in recent years this has proved problematic due to an increase in red mite populations. These pests will live under the felt, but only if you allow them to get out of control. Bituminised corrugated roofing sheets are a very viable alternative to felt, but it is a good idea to place a layer of small gauge wire mesh underneath them to stop rats and other animals gaining access via the corrugations.

Pop hole

All houses should be fitted with a pop hole. This is the name given to the small hen-sized door that allows the hens access to the outside. If you plan to keep very large breeds you may need to get the size of the pop hole altered to accommodate their extra bulk. The best type of door to have on the pop hole is a simple drop-down type. Those made to

Left: The wire mesh under this corrugated roof will prevent rats or stoats gaining access to the house. Use small gauge mesh to be sure.

Above and right: Clear corrugated roofing can be used over chicken runs to make them biosecure and still provide light. Green mesh provides shade from full sun.

Left: *The vertical drop-down type of pop hole on this house is easily opened and closed by way of a string that leads to the outside of the house. Note the mesh-covered ventilation holes near the top of the dividing wall between the house and the run.*

hinge upwards can become clogged with shavings and those that slide at an angle of 45° may start to stick after a period of use. A door that drops down vertically is less prone to these problems.

Your poultry rely on you to protect them as best you can. One important responsibility is remembering to shut the pop hole of the house each night so that the birds are safe and secure.

Electronic devices have arrived on the market that automatically shut the pop hole for you. A timing device slowly lowers the door at night after the hens have gone to bed and opens it in the morning to let them out. These devices activate either by monitoring the available daylight or at a preset time.

Recent reports indicate that very overcast weather can seriously reduce the amount of daylight available and can cause light-controlled devices to shut early, leaving the hens outside. The best designs incorporate a clock timer. You need to alter the closing time to suit the current daylight hours, but

the timer makes the device immune to early closing. They are proving very popular, especially during the winter months when many people are unable to get home from work until after dark, by which time the local fox is already on its evening prowl.

Perches

In the wild, jungle fowl automatically fly up to roost at night and this trait has passed down to its descendant, the domestic hen. All hens will take to perching, but some take quite a bit longer to get the idea fixed into their brain than others.

Perches should be removable for ease of cleaning. They should have rounded edges, with no sharp corners, to enable the birds to grip them easily. A piece of wood measuring 5x5cm, with the sharp edges planed off, is very good for this purpose.

The height of the perch from the ground will depend on the breed kept. Heavy breeds, such as Orpingtons, Cochins and Indian Game, need lower

Above: Placing old newspapers below the shavings can make cleaning the floor of the house easier.

perches, set at approximately the height of the bird's back; otherwise leg and foot injury may occur when they jump down in the morning. For lighter breeds and hybrid hens a perch set at around 60cm from the floor should be suitable.

Perches should be set higher than the nest boxes to encourage the birds to roost, otherwise you may find hens sleeping in the nest box and fouling it. You can have more than one perch in a house, but make sure that they are all set at the same height or there may very well be squabbles amongst the birds for access to the highest perch. In general, allow around 20cm of perch space for each bird. This gives them a chance to snuggle up close or move apart, depending on the temperature. Allow slightly more perch space if the birds are large or belong to a breed with feathered feet. These will need the extra space to prevent them breaking their foot feathers when they roost.

Droppings boards

Droppings boards seem to have gone out of fashion in recent years, but they are a very useful addition to a poultry house. Birds produce most of their droppings at night whilst roosting. Placing a droppings board, a tray or a thick piece of plastic under the perches to collect these droppings makes cleaning out the house a much easier task.

Nest boxes

Hens will quite happily share nest boxes and do not need one each. Allow one nest box for every four birds. Nest boxes usually measure 30x30cm.

Often, nest boxes are situated so that egg collecting can be done from outside the hen house. This is a very good idea, but weather protection can sometimes prove poor at the point where the box joins the house. Keep watch to ensure that nest boxes remain clean and dry.

If you use free-standing nest boxes inside the house, locate the entrance to the nest box in a dark

Below: The two nest boxes fitted to this house are very easy to access from the outside, making both egg collection and cleaning a simple matter.

Right: Chickens very quickly learn to walk up a ramp or small ladder to gain access to their poultry house.

part of the house away from direct light. This will encourage the birds to use the box and helps to prevent egg eating (see page 65).

Runs

Even if you decide to free range your birds, there are times when you need to keep them enclosed. This might be when visitors arrive, when you are medicating the birds or if there is a problem fox in the area. In the event of an outbreak of avian

A typical poultry house

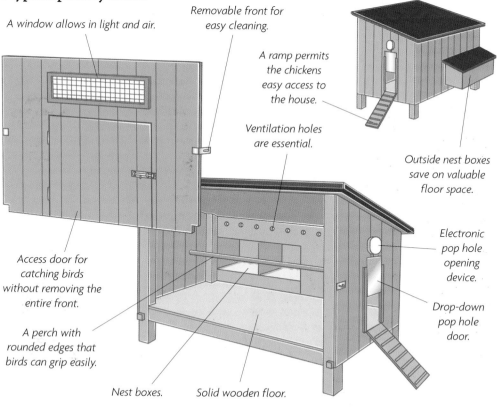

A window allows in light and air.

Removable front for easy cleaning.

A ramp permits the chickens easy access to the house.

Ventilation holes are essential.

Outside nest boxes save on valuable floor space.

Access door for catching birds without removing the entire front.

Electronic pop hole opening device.

A perch with rounded edges that birds can grip easily.

Drop-down pop hole door.

Nest boxes.

Solid wooden floor.

influenza, all poultry keepers in the outbreak area may be obliged by law to house birds indoors for a few weeks.

You should therefore consider how best to keep your birds under cover if the need arises and take this into account before buying a house. If you choose a free-standing house, it is a good investment to buy a small enclosed run to go with it that can be covered with a tarpaulin roof in the event of an emergency.

Other considerations

Remember to consider the welfare of the person who will be looking after the birds. If you (or they) have difficulty bending or have back problems,

then cleaning out a small house low to the ground may be awkward. You can now buy houses that can be tended with little or no bending. If you lack strength, you may not be able to move a portable house around the garden on a regular basis without assistance, so a permanent pen with a fixed run may prove a better option. The choice of house should suit both you and your hens. That way you will both be contented.

Siting the house

Where you site the house will usually depend on the layout of your garden. It is not a good idea to site it where it will be an annoyance or inconvenience to your neighbours or yourself. Equally, there is no need to hide it away, as if keeping poultry were an unsavoury habit! A well-built, attractive house could even become a feature and talking point of your garden.

If you love your lawn do not site your chicken run on it. Hens will not concern themselves one jot that scratching a deep dust-bathing area in your carefully tended lawn might upset you. They can reduce a neat lawn into something resembling a lunar landscape if the whim takes them.

Below: *The outward-facing apron of wire on this portable run prevents foxes digging underneath it.*

Above: *This house and run has a covered area at one end. This keeps the feeding area dry and allows the birds somewhere to shelter from hot or wet weather.*

A removable cover provides some shade during hot weather.

Moving runs

Move runs regularly to keep the ground in good condition.

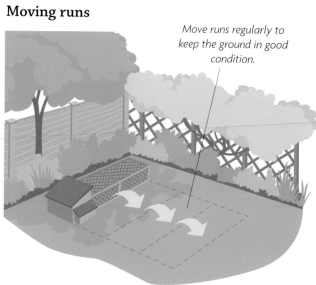

Above: *Some runs have handles at each end and can be fitted with wheels to make it easy to move them around.*

Left: *Any bare patches of ground will soon recover if allowed a period of rest before reuse.*

A movable house

An area of rough grass, with the shelter of some trees, is ideal for a movable house with an attached run. Move the house to fresh ground on a regular basis to prevent the ground becoming soiled. This can be done on a rotation system, whereby you move the run back to its original location once the grass has recovered.

Always try to site the house with its back to the prevailing wind. This will prevent rain blowing in through the pop hole or any front-facing windows. The roof at the back of the house is invariably lower, thus lessening the impact of the wind.

Chickens much prefer living with some form of shelter and protection from both the sun and the elements. They do not generally like wide open spaces. If no shrubs are available then try to provide a trellis or fence.

A run in a permanent location

A permanent run should be as large as you can possibly make it. There is no maximum size and the closest you can get to a predator-proof 'free range' environment the better. Overcrowding amongst the birds will cause them stress, which in turn can lead to disease and vices. A recommended stocking density for a permanently sited chicken run is 10m^2 of outdoor run space per Large Fowl. Thus, a static run measuring 6x5m would be suitable for three Large Fowl on a permanent basis. If space is at a premium, it is advisable either to keep fewer birds or keep the smaller Bantam breeds that require less space.

If you are using an open-topped run, trees and shrubs, whilst providing shelter, may also offer tempting roosting places for your birds, especially if the branches actually overhang the poultry run. Trim back overhanging branches so that the birds cannot fly directly up to them, and thence out of the run. Locate the house sufficiently far away from the sides of the run to prevent it being used as a launch pad to freedom. In addition, clipping the primary flight feathers of one wing will help prevent birds from flying over the wire.

Run flooring in wet weather

A muddy, wet run is not pleasant either for you or your hens. Even if you have carefully built it to the correct size, a permanent run can start to look very stale and sorry after a prolonged downpour. A muddy run can lead to soiled eggs, foot problems and disease; at best you will end up with very dirty and miserable birds.

If the run area is not sufficiently large to maintain permanent grass, then you should seriously consider covering all or part of it with a material that will keep it draining easily and free of mud in the winter months. The best materials for this purpose are either shingle, gravel, or wood chips laid down to a depth of 10cm. Wood chips are an ideal cover, keeping the run floor dry and the birds happy. Best of all they are easy to obtain from tree surgeons at a very reasonable cost; some even provide them free of charge. Do not confuse wood chips with either wood shavings (see page 28) or the bark chippings sold at garden centres for mulching plants. The latter can carry a fungus that is dangerous to chickens and should not be used.

Above: Chickens will take to roosting in trees if the opportunity arises.

Above: This well-constructed run has been kept in good condition by keeping the stocking levels correct.

Right: Hens enjoy finding a shady place to shelter from hot sun. Planting suitable shrubs will provide this.

Enhancing the environment

If you choose to confine your birds to a permanent run, it is worth making that run as interesting as possible for your birds; bored chickens can develop vices. Try hanging up some things for them to peck at. A shiny CD suspended by a string or a wire peanut feeder will keep them busy. A partly open bale of straw with some wheat scattered inside will produce many hours of interest, as well as providing some additional exercise.

Planting tough, hardy shrubs and small trees will provide a shaded area for the chickens to rest under on hot days and will be much appreciated. A large pampas grass is tough and able to withstand living with hens. Holly and conifers should survive, as will bamboo. Allow the plants to become very well established by fencing them inside a small area of the run until they are big enough to be chicken-proof. Do not plant smaller soft plants, such as flowering annuals, as these will be trampled and destroyed very quickly.

Chickens will not generally eat plants that are dangerous to them. However it is best to avoid planting yew, cherry laurel, privet or laburnum, as young, inexperienced birds might be curious enough to peck at things they find interesting until they learn better.

Right: This young cockerel is fascinated by his own reflection. A shatterproof mirror or suspended CDs will provide activity and interest for your chickens.

Above: When roosting, the feet of the birds are protected from frost by their feathers.

Predator-proofing the run

The best-known predator of poultry is the fox. Even if you have not seen it, there is probably a fox in your neighbourhood, more than eager to check out the arrival of any chickens that appear on its patch.

If you are free ranging your birds, it does not mean they will always be safe from attack just because you lock them up when it gets dark. It is most important for budding poultry keepers to note that foxes are not just nocturnal animals. Many fox attacks on poultry happen during daylight hours. This is especially true in the spring when the foxes have cubs to feed, and again in late summer when they are teaching the cubs to hunt. Other predators will also attack poultry, as will stray dogs, but the foxes' reputation as the number one chicken-killer is well deserved.

Installing a fox-proof fence

Although free ranging is the ideal management for hens, be aware that losses will occur at some time unless you provide a physical barrier between the birds and the fox. Foxes are very determined creatures and can scale far higher fences than most people would suppose. They can also dig and use

Deterring human predators

If you are keeping breeds such as Asian gamefowl, it is important to be aware that human predators may well prove as troublesome as the fox. Wireless infrared alarms are now readily available from hardware shops or the internet. These can be set to protect doorways and gateways and can prove a good investment.

their teeth to rip at chicken wire. Luckily, there are ways to prevent them eating your hens.

For a permanently sited chicken run, the most cost-effective method of fox-proofing is to install a 2m tall, 13x13mm 19g chicken wire or weld-mesh fence, with single-strand electric fencing wire at the top and bottom of the fence. Site the bottom electric fence wire at the base of the run about 10cm high and an equal distance away. The top electric fence wire should form an overhang and also be offset by about 10cm (see diagram)

Any outward-facing overhang, electrified or not, at the top of the fence, will help discourage climbing foxes.

To deter a digging fox at the bottom of the fence, bury the chicken wire underground to a depth of 45cm. Alternatively, make a wire 'apron' that faces outwards for 45cm all around the run. This can be covered with grass, turf or paving slabs.

Creating a solid barrier

Often, foxes will only make a concerted effort to gain access to a chicken run if they can see or hear the birds on the inside of the fence. Thus, a careless hen that decides to sleep next to the chicken wire would drive many foxes into making a huge effort to dig under, climb over or rip through the wire. A hen that is safe inside a house and not visible would be less of a target. For this reason, an additional defence against the fox is a 60cm-high solid barrier all around the base of the run to prevent the fox looking directly into it.

Electric fencing

Electric fencing, in the form of specially designed netting, is proving very popular for larger areas.

Electric fence units deliver a powerful, but not fatal, shock to anything that touches them. One drawback is that you constantly have to be aware of and remove vegetation growing up around the netting and shorting it out.

To overcome this problem you can either use a weedkiller beneath the fence or run a narrow strip of plastic directly under the netting to restrict weed growth. Damp-proof coursing material works well for this purpose. Do not make it too wide as it will insulate the fox if it is treading on it. A determined fox may decide to jump the fence, but most will stay away once they have received an initial shock. Even with an electric netting perimeter fence, the houses must still be shut up each evening. Make this part of your daily routine.

A predator-proof fence

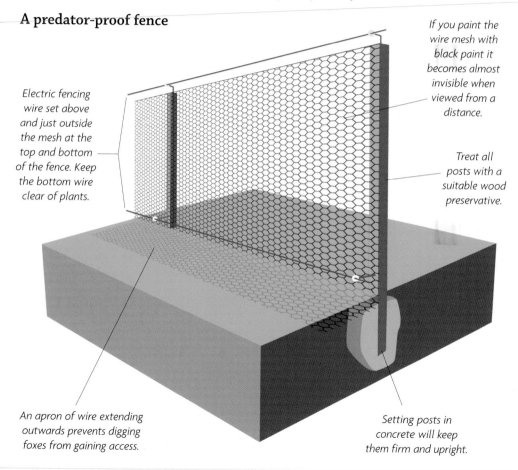

Electric fencing wire set above and just outside the mesh at the top and bottom of the fence. Keep the bottom wire clear of plants.

If you paint the wire mesh with black paint it becomes almost invisible when viewed from a distance.

Treat all posts with a suitable wood preservative.

An apron of wire extending outwards prevents digging foxes from gaining access.

Setting posts in concrete will keep them firm and upright.

Management and feeding

Here we look at the day-to-day care of your poultry. For advice on the specialist management and additional feeding requirements of breeding stock, chicks and growers, see *Breeding* (pages 42-61).

Wood shavings make a good covering for the floor.

Chopped straw will compost more easily than shavings.

the advantage of rotting down quickly, making a wonderful compost. Chopped straw is also suitable, but avoid using hay for bedding as it can contain fungal spores that may cause serious respiratory problems if it gets damp. It is possible to use some hay in the nest boxes provided it is replaced often. Hens certainly like nesting in hay and if a hen tends to lay her eggs on the floor, it may encourage her to use the nest box instead.

The art of keeping a flock of adult chickens healthy, contented and laying eggs is, thankfully, quite straightforward. Most good poultry husbandry is simply commonsense. You do not have to fuss over your birds as if they were breakable porcelain, but neither should you be so lax that you miss vital signs when things are wrong.

Good stockmen spend a lot of time just watching their animals. Apart from being a good excuse for a relaxing break, this will tell you more about the health and well-being of your birds than almost anything else. You will soon get to know how they react and behave in different situations. Once you learn that, you will very quickly spot when things are different from the normal routine.

You will need to tend your birds daily, letting them out in the morning and shutting them up safely at night. They will require feeding, watering and a visual check-over. You must also collect any eggs that have been laid.

Bedding

Clean white wood shavings are a suitable bedding for both the poultry house and the nest boxes. Hemp-based products are also proving very popular, as they are more absorbent than shavings and have

The dust bath

Dustbathing is a natural instinct for poultry and helps to keep them free of parasites. You can provide a permanent dust bath. A covered dust-bathing area that they can use all year round will meet many of their natural requirements. This can consist simply of four posts knocked into the ground to a height of 60cm with a roof over it.

Below: Dustbathing helps to remove parasites and also provides an enjoyable pastime on warm sunny days. This chicken has excavated its own dust bath.

Introducing new birds

Introduce your hens into their roosting house at night, so that when they come out into the run in the morning for breakfast they will already associate the house with where they sleep.

Even if you plan on free ranging your birds, keep them confined to the house for a couple of days so that they learn where they are to roost. When you first let them out, do not hurry them, but allow them to exit the house at their own pace. Confine them to an area around the house for the first day

Above: *This old ceramic sink filled with a mixture of silver sand and sieved dry soil (plus some mite powder) has been provided for use as a dust bath, much to the obvious delight of this hen.*

Loosen up the soil under it and mix this with some dry sand or ashes to get it started and leave it for the hens to do the rest. If you are using a movable run, you can provide a box filled with dry sand, ashes or dry earth. Adding a dash of diatomaceous earth to the mix enhances the parasite-removal value of the dust bath.

When on free range, hens will decide for themselves where to dust bathe. As this may be on top of your newly planted flowerbed, providing tempting alternatives is always a good idea.

Quarantining new birds

Once you have acquired your birds and brought them home, do not introduce them to their house straight away. To avoid introducing potential diseases to an existing flock, always quarantine new birds in a separate pen, away from the resident birds, for at least two weeks. If any parasites have hitched a ride home with the new birds, treating them now with a suitable mite spray will help to keep the hen house clear of these pests for a good while.

Above: *The hunched and ruffled appearance of the bird at the back of this photograph is a sure sign that it is unwell. These birds are in quarantine.*

Below: *Introduce new birds (left) to existing flocks (right) after they have been able to observe each other from the safety of a wire pen for a few days.*

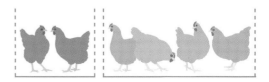

to be sure that they go into the house that night. After that they can be allowed to range freely and will automatically go to bed as dusk falls.

When introducing new birds to an existing flock, be aware that a single lone hen will be bullied by the resident hens, so it is always best to introduce two or more together. Confine the newcomers to a temporary pen where they can meet the existing residents through the safety of some wire for a few days before allowing them to mix.

Cleaning the hen house

Clean out the hen house once a week, making sure you have the right tools for the job, including a short-handled shovel. A hand-held hoe is very useful

Cleaning an Eglu

Left: Eglus can be disassembled for easy cleaning. Simply pull out the dropping tray to empty it.

Right: If you wash the tray, make sure it is completely dry before replacing it.

Weekly cleaning

1 A good-sized doorway makes cleaning a poultry house easy. Be sure to remove old litter from the corners using a brush and hoe.

2 Sprinkling disinfectant powder will help keep the house hygienic, but may hinder the composting ability of the bedding.

3 Replace the bedding. Provide a good layer, especially in the winter months. The bedding being used here is chopped miscanthus.

Handling a chicken

When picking up a chicken, use one hand over the top of the bird to stop the wings from flapping and place your other hand underneath the bird, with your middle finger between the bird's legs. Use the fingers on either side to hold the bird's legs firmly to stop it struggling. You can use your thumb and little finger to hold the tips of the bird's wings still. Resting the bird's breastbone on your forearm and close to your body will make it feel more secure and less prone to wriggle.

Above: *A well-held bird will feel secure and should not struggle. Keep it close to your body for support.*

Right: *Placing your middle fingers between the bird's legs allows you to hold it securely. The legs will be gripped between your outer fingers.*

Above: *Placing your hands over the chicken's back will restrain its wings and stop the bird flapping. Keep a firm but gentle hold as you lift it.*

for reaching into corners, as is an old flat paint-scraper. If you have provided a droppings board in the house (see page 20), clearing this daily will make the weekly clean much easier. Use a wheelbarrow or large bucket to take away the droppings. If you have more than one hen house, rinse or spray the tools between use to prevent spreading parasites between the houses.

Checking for parasites

Check the poultry house often for parasites such as red mite. These build up very rapidly and will cause severe problems amongst your stock. (See *Health care* pages 68-72).

Get used to picking up your birds and handling them on a regular basis. Hens have poor night-time vision and this makes it easy to catch them at night after they have roosted. If done on a regular basis, you will very soon realise if a bird is losing weight, which may hint that it is unwell.

Moulting

In late summer, birds start to moult their feathers, meaning they shed the old feathers and grow new ones. At this time, the birds stop laying and start to

Left: Gaps will appear in the wings where the old feathers have fallen out.

Below: These frayed feathers will soon be replaced by new ones.

look very tatty. Do not cut back on their feed, as they need good nutrition and plenty of protein to grow new feathers ready for the coming winter.

Ready-mixed and fresh foods

Today, ready-mixed poultry feeds are widely available. These compound feeds are designed to meet the chicken's dietary requirements at every stage of its life and should be the mainstay of its diet. Adult hens need a layers ration, which provides around 16% protein. Pellets tend to be less wasteful than dry mash and it is easier to ensure that no food is left in the bottom of the trough.

In the early evening, you can scatter a grain feed, such as wheat, on the ground. Avoid feeding grain in the mornings, otherwise the birds will not consume enough pellets and the overall nutrients in the diet will drop, as will egg numbers. An average-size, active hen eats about 130gm of feed a day, plus about 20gm of grain feed. In winter, increase the evening grain feed to about 30gm. As a guide, a half-litre jug holds approximately 375gm of chicken pellets. Birds should always go to bed with a full crop to enable them to sleep soundly. Overnight the crop slowly empties as the birds digest the food stored in it.

Left: This Buff Sussex cock is in full moult and has already lost many body feathers, especially from his breast and thighs.

Above: *Any occasional treats, such as rice, pasta or broccoli, should not contain salt.*

Above: *These hens are enjoying a feed of wheat in the garden as evening approaches, which is the best time for doing this.*

Below: *Hens appreciate a piece of apple as a treat. The tamest hens will always approach first.*

Layers pellets contain all the nutrition a laying hen requires.

Good-quality, clean wheat is best fed as an evening meal.

Left: *Hand feeding wheat or mixed corn encourages your hens to become friendly.*

Provide your chickens with fresh greens and vegetables on a regular basis. Hanging these up will prevent them becoming soiled. You can offer treats, providing they do not contain salt or meat, but use these only very sparingly.

Hens like routine and it is best to feed them at a set time each day. You may decide to feed the birds twice daily or provide a hopper so that they can help themselves freely throughout the day; both methods have their merits.

Keep all feed under cover to avoid contamination by wild birds. It is never a good idea to leave feed out overnight, as this attracts rats. Store feed in a rat-proof container; a metal dustbin with a well-fitting lid is ideal.

Water

It should go without saying that fresh water must be available at all times. It is best to use a poultry water fount rather than an open bowl. Raising it off the ground on some bricks will help it stay clean. Chickens prefer cold water, so site the water in the shade. Always empty out any old water before refilling the drinker with a fresh supply.

Below: Suspending a food hopper off the ground by hanging it from a chain helps prevent the feed becoming soiled or damp.

Left: If water founts (here galvanised metal) are raised off the ground, the hens will not scratch dirt into the water.

Grit

Because hens do not have teeth, they need to grind up food in their gizzard. A supply of hard flint grit assists the grinding process. Modern compound feeds should provide sufficient calcium in the diet, but providing additional oyster shell grit helps to keep the eggshells strong. Supply the grit in small open containers so that the birds can use it when needed.

Keeping everything clean

Wash feeding troughs and water founts regularly, as they rapidly become soiled and contaminated with stale food and faeces. There is no 'best type' of feed trough and many types of heavy duty bowls are perfectly suitable. For birds that are feeding throughout the day, a purpose-built hopper provides a reservoir of food without the birds being able to scatter it all around.

Left: Make sure that a supply of fresh water is available to your birds at all times. These suspended plastic water founts are an ideal way of doing this.

Washing a water fount

1 In summer, light causes algae to grow rapidly on the insides of a plastic water fount.

2 Use a brush or your fingers to get well into the corners when cleaning the fount.

3 Make sure that all the algae is removed and then rinse thoroughly in fresh water.

Above: A clean fount ready for use again.

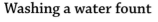

Eggs

Their propensity to lay a regular supply of eggs has made poultry one of the creatures most frequently found on smallholdings and farms throughout the world. Not only do they lay eggs, which can be eaten as food, but by incubating fertile eggs, poultry are able to provide their own replacements.

The development of the egg

The time from ovulation until an egg is laid is about 25 hours. As the yolk travels from the ovary along the reproductive system, it becomes surrounded in membranes, albumen (the egg white) and finally

Inside an egg

Tiny pores in the shell allow gases and water through.

Albumen (egg white) is a support medium for the yolk.

The airspace provides air for the chick as it chips its way out of the shell.

The yolk provides nutrients for the embryo, which develops from a germinal disc of cells on the yolk surface.

The chalazae hold the yolk in the centre of the shell.

Above: *The airspace at the blunt end of an egg increases as the egg gets older. This provides a way of checking freshness – stale eggs float in water.*

shell. The rotation of the yolk produces the two strong cords (called chalazae) found on each side of the yolk. These suspend the yolk in the middle of the egg.

If the egg has been fertilised, the yolk will act as a food supply for the growing chick and as such it is very rich in nutrients. The group of cells that would become a chick is called the germinal disc. This can be seen as a very small white dot on the surface of the yolk. In a fertilised egg, the edges of this dot become very smooth and uniform, and the shape is often compared to that of a doughnut.

If you open an egg and look at the blunt end you will see a small airspace. In a fresh egg it is about 2cm in diameter and set slightly to one side. Its purpose is to provide a growing chick with a supply

of air during the period when it is starting to chip its way out of the egg. This airspace increases in size as the egg ages due to the evaporation of the contents. The size of the airspace is a good guide to the age of an egg. When it becomes very large, the airspace causes a stale egg to float when placed in water.

The eggshell

The shell of a hen's egg is composed of 94% calcium, and high levels of usable calcium are needed to keep the shells strong. The hen provides about 2.5g of calcium for each egg she lays. The shell is covered with up to 17,000 tiny pores that allow oxygen, carbon dioxide and water to pass into and out of the egg.

Egg shells have a natural 'bloom' that helps to protect them from environmental contamination. If an egg is washed to remove dirt, this natural protection is lost. Therefore, keep the nest box material clean and collect the eggs as soon as possible after they have been laid to keep washing

Above: *Eggs are found in many colours and sizes. It is the breed that determines the colour of the egg. Older hens may lay paler shades than young birds.*

to a minimum. If you must wash eggs, always use water that is warmer than the egg, as this helps stop bacteria being drawn into the shell.

Shell colour

Shell colour is hereditary and the breed of the bird defines the colour of the eggs the hens will lay. Colour is determined by two genes, brown and blue. When both these genes are absent, the shell colour will be white. Modifying genes create the different shades of colour. Over the years different breeds have been selectively crossed, with the result that we now have various breeds that lay an assortment of egg colours.

The Araucana, a breed from Chile, lays a blue egg and, unusually, the colouring runs right the way through the shell. In all other breeds, the colour is only the top coating and can be scratched off. The blue eggs found in supermarkets today owe their origins to the Araucana.

Speckled eggs can crop up in a few breeds but it is a very limited trait, usually confined within certain strains. This is a pity, as a nice speckled egg looks really good at the breakfast table.

The depth of shell colour pigment will depend on the laying qualities of the hen, with older hens frequently laying paler eggs than younger

Left: *Breeds such as the Araucana lay blue eggs.*

ones. Pigment colour can also be affected if the birds become stressed, for example as a result of environmental disturbances, internal and external parasites, illness or the presence of nearby predators.

Daylight and eggs

A chicken's brain contains light-sensitive cells, known as extra-retinal photoreceptors. These cells help control the egglaying process. When the cells receive sufficient long-wavelength light, they send a message to the ovary, which springs into action and egg production starts. The increasing daylight hours in spring will bring the hens into lay, but as the winter months approach, the shorter days will cause a slow down in egg production. On average, a hen needs 14 to 16 hours of light on a regular basis to stay in lay. This can be either natural light or a mixture of natural and artificial light.

Most traditional breeds will stop laying at some period over the winter months unless provided with additional artificial lighting. If you decide to

Artificial lighting

■ Night   ■ Day   ■ Artificial light

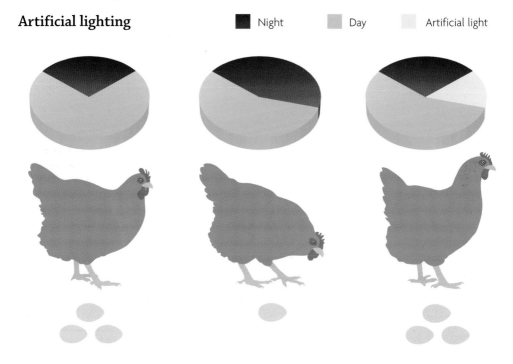

During the long summer days hens lay more eggs. Up to 16 hours of daylight are needed for consistent laying.

As winter arrives the darker nights will mean fewer eggs are laid.

Artificial light can be used to increase egg production again. Add this in the morning period.

provide your birds with extra hours of artificial light in winter, do so in the early morning, rather than at night. This allows the hens to go to roost as usual at dusk and eliminates the chance that the light will suddenly switch off, leaving them sitting on the floor in the dark. If you keep a cockerel, providing early morning lighting may not be popular with your neighbours, as he will wake up and start crowing as soon as the lights go on.

To remain in good health it is vital that hens take a break from laying. For this reason you should not provide additional lighting when the birds go into the yearly moult. A hen that is provided with near-permanent daylight will not live to a ripe old age.

New laying hens

It is in the nature of hens to dislike changes to their environment or circumstances. Any change, even a small one that you would class as an improvement

Above: *The small egg at the left is a Marans 'wind egg' shown next to a Bantam Araucana (blue) egg and a pullet Marans egg for size comparison.*

to their situation, may stress them and cause them to stop laying for a while. Therefore, do not expect any newly acquired hens to lay eggs straight away. You might get a couple of eggs that were already on their way through the reproductive system, but after that the hens will take a short break. Once they adapt to the change they will come back into lay.

When a hen is just starting out on her laying career, she occasionally lays very tiny 'wind eggs'. They can frequently be as small as a marble and usually have no yolk. They are the result of a small leak of albumen into the oviduct. The egg-producing mechanism responds by covering it with a shell, just as it would do with any normal egg. Wind eggs are rarely found once the hen starts laying properly.

Egg shape

The shape of the egg is formed by the structure of the uterine wall. If eggs take on a ridged or striated appearance, check that the hens have not become over-fat, as internal fat pressing on the reproductive

Left: *As she matures towards laying, the face of a pullet will begin to change to a deeper red colour, as shown in this Fayoumi.*

organs can be one cause of this type of egg. Abnormal eggs can also result if hens have been suffering from a respiratory infection. This will often correct itself once the bird is well again, although these diseases can sometimes have a permanent effect on the shape and colour of future eggs.

Soft-shelled eggs

Soft-shelled eggs can be very startling to a novice poultry keeper. They look like normal eggs, but when you pick them up you find that there is no

Above: The egg from this Appenzeller is quite large compared to the body size of the bird. This bird is clearly used to being handled by its owner.

strong shell surrounding them, just a transparent membrane that is easily broken. If hens start producing soft-shelled eggs, make sure they are getting enough dietary calcium. Concentrated calcium and mineral supplements, especially formulated for laying poultry, are very effective at improving shell quality.

Double yolks

A double-yolk egg results when two yolks are released from the hen's ovary at the same time. A young hen that is just getting into her laying stride often produces these, as will older birds who are only laying intermittently. Double-yolk eggs will not hatch as there is not sufficient room for two chicks to grow inside the single egg shell. You can often spot a double-yolk egg by a small ridge that can be felt around the middle of the shell.

Blood, or meat, spots in eggs

Contrary to popular belief these do not indicate that an egg is fertile. They are caused either when a small blood vessel in the ovary ruptures as the yolk is released into the oviduct or when a small piece of lining from the oviduct is included in the egg.

Pale or odd-coloured yolks

Yolk colour depends on what the hens have been eating. Hens on free range will ingest enough plant material containing the required carotene to keep the yolks bright yellow. To improve the colour of egg yolks, try adding some marigold petals, carrots, kale, broccoli, chard or pumpkin to the hens' diet.

Hens that have been feeding on the pasture weed shepherds purse *(Capsella bursa-pastoris)* or on acorns can lay eggs with unpleasant-looking green yolks. Needless to say, these eggs are far from aesthetically pleasing when cracked into a frying pan. They are best discarded until the hens are back to laying normal-coloured yolks again.

Above: *It is often fairly easy to guess which eggs will have a double yolk. Here are single and double-yolk Buff Sussex eggs.*

Right: *Clear proof that the larger egg shown above did indeed contain two yolks.*

Breeding chickens

Breeding and raising poultry can be great fun, and watching the chicks grow and develop into adults is very rewarding. However, before you embark on this journey do bear one thing in mind. It is a sad fact of life that every year forlorn advertisements appear all over the country, desperately asking for good homes for 'sweet', 'pretty', but very much unwanted cockerels – advertisements that have exceedingly little chance of finding a taker.

About 50% of each batch of chicks hatched will be cockerels and even top show breeders have problems finding homes for truly excellent specimens of the breed. Cockerels are noisy and quarrelsome and are generally unwanted by anyone except breeders – and breeders usually have more than enough cockerels of their own. The most humane thing is to cull any you do not wish to keep yourself, as soon as possible before you become over-fond of them or they start developing 'personalities'. Alternatively, you may decide to raise them as an ethical form of meat for your own table.

Selecting the best breeding stock

Once you have decided to breed from your birds you must select the breeding stock. Whatever sort of birds you keep they should be good representatives of the breed. Unless you have a specific reason for doing otherwise, it best to breed only from pure breeds. Surplus pullets (females up to one year old, see also page 58) will find a ready market and you can be reasonably sure of what temperament and utility qualities to expect. Remember that it costs no more to feed and raise good-quality examples of stock than it does poor ones.

As with all breeding, if you find any defects on one side of the breeding line try to counteract them

Hens will need to produce good-shaped eggs.

> ### *Ethical produce*
>
> *It is far more ethical to know how and where your meat was raised than to buy an intensively reared bird from a supermarket. Do not allow cockerels to suffer the stress of local markets and livestock dealers, only then to suffer the same fate that you could have provided humanely at home.*
>
> *It is best to ask an experienced poultry keeper to show you the quickest and most humane way to dispatch birds. It is not something that can really be learnt from books without the chances of something going wrong. If you wish to breed poultry, knowing how to dispatch birds humanely is all part of that process.*

by adding excellent points on the other side. Thus, if your Light Sussex cock bird is a bit poor in his hackle markings but otherwise excellent in type, pair him to hens that excel in their hackle markings. Hopefully, this will produce a percentage of chicks that carry the better hackle marking trait.

Birds used for breeding should be healthy and fully adult. It is best to use hens over one year old, as they will have proved themselves as productive, vigorous birds.

The cock bird should be at least six months old, although some more precocious Light Breed males are often fertile and active when slightly younger. Very Heavy Breed males, such as Orpingtons and Brahmas, can often become infertile after the age of two, whereas in the lighter breeds males can continue fathering chicks for a number of years. Hens may continue to be fertile up to about six years, but the number of eggs they produce at that age will be few.

The male will be the father of all your future chicks.

Left: *A breeding trio, here Cream Legbars, should be well matched and showing all the signs of good health and vigour.*

To ensure that the eggs are fertilised by the correct male, pen the birds together and away from any other cockerels. It is amazing how determined a cockerel can be, and even pairings that you may have considered physically impossible due to vast differences in the physical size of the birds can happen. The proof of this mismatch only comes to light when you discover strange-looking cross-bred chicks emerging from your carefully incubated eggs.

The usual number of birds in a breeding pen is known as a trio, consisting of two hens and one cockerel. Light Breed males can cope with more hens (about six is not uncommon), whereas the heavier breeds do best with no more than three or four hens per male to ensure optimum fertility.

Providing the correct diet

Breeding birds need good-quality nutrition in order to supply the egg with all the nutrients that the developing chick requires. Therefore, it is best to switch the birds from a layers ration to a special 'breeders ration' a few weeks before the eggs are needed. This food has a higher protein content than layers ration, plus extra vitamins, minerals and amino acids to provide for a healthy embryo.

Mating

The act of mating in poultry is referred to as treading, because the male mounts the hen by standing on her back whilst gripping her head and neck feathers with his beak to aid his balance.

Ensuring fertility

To get good fertility from your chickens it is best not to use too many hens in a breeding pen. The classic arrangement of a trio, one cockerel with two hens, can be modified as shown here.

It is usual to run Heavy Breed males with only a few hens at a time.

The more active Light Breed males can cover a larger number of hens.

Below: *Hens can suffer from quite severe feather loss during the breeding season. The back of this hen shows where the cockerel has been 'treading' her.*

Above: *It is easy to see here which hen is the cockerel's favourite; one shows the feather damage of mating while the other hen is in good condition.*

It is not uncommon for a male to decide on a favourite hen and constantly mate with her to the exclusion of the other hens, resulting in poor overall fertility. You can usually spot the favourite, as all the other hens will have immaculate feathering, while the one poor favourite will have a harassed look and large bare patches on her back from the overzealous attentions of the male. If this happens it is best to remove the favourite from the breeding pen every second day, so that he will mate with the other hens as well.

A hen's back and sides can be damaged during mating. This is mostly caused by the cockerel's sharp toenails, which can gash the hen's sides and back. It may also result from spur damage, although this is less common. If a hen is starting to lose too many feathers from her back, you may wish to fit a poultry saddle to protect her. These are made of strong cloth and fit around the wings and over the back of the hen. They have a ridge of strong cording to permit the male to grip when treading, but will give a level of protection to the hen's back.

Collecting and hatching eggs

In theory, hens can be fertile two to three days after their first mating, but it usually takes about a week before you can start to collect fertile eggs for setting. If the hens have been running with another cockerel previously, wait a minimum of two weeks before collecting any eggs, as they are very likely to be fertile from him instead of the male that you have selected.

Eggs set for incubation must be clean; it is no use setting dirty eggs, so do make sure that the nest box material is very clean and fresh. Washing eggs before incubation is best avoided if at all possible (see also page 37). Specialist egg-washing products are available, but you will get a much better result by keeping the eggs clean and unsoiled in the first place. Collect hatching eggs at least twice a day to reduce the chances of them becoming soiled or damaged. Check the eggs over and do not set any that are cracked, misshapen or damaged. A chick will have a better chance of hatching from an egg with a good shape and a strong shell.

Contrary to popular belief, you should never keep eggs warm while gathering together a clutch ready for incubation. Eggs do not start to incubate until they have been kept at a constant body heat for 24 hours. They need to remain cool to prevent any premature development until you want to set them.

Store the eggs in a cool place – but not in the fridge! A temperature of 12°C is best and a cool pantry or cellar is ideal. Experiments have shown that is it best to store pre-incubated eggs intended for hatching point upwards. Try only to set eggs that are under a week old, as these will give the best hatching results. If you need to keep them a little longer before setting, turn the eggs once every day to prevent the contents sticking to the shell. Do this by tilting the egg storage box at a 45° angle one day and then tilt it back 45° the other way the following day. This way the eggs are resting at a different angle each day. Even with turning, the viability of the eggs will start to drop off rapidly after 11 days.

Left: Keep the nest boxes clean and collect the eggs often. If you want to set the eggs for incubation, keep them cool before you do this.

Providing the correct environmental conditions for a successful hatch

Eggs require a constant temperature of 37.7°C. They will tolerate a very minor fluctuation in this temperature, but bear in mind that a slightly hot incubator is far more damaging to the eggs than a slightly cool one. Hatching eggs require a relative humidity (RH) of about 60% for the first 18 days of incubation and at least 80% RH during the hatching process. Eggs also need ventilation so that gasses, such as oxygen and carbon dioxide, can be exchanged through the porous egg shell. The eggs also need turning, so that the embryo does not stay in one place and stick to the shell.

A broody hen satisfies all these requirements quite naturally. If the weather is too hot she will not sit as close to the eggs, if cold she will press down on them to keep them warmer. She will shuffle the eggs around into different positions using her

feet or beak. In addition, the oil from her feathers coats the eggs with a protective covering that helps keep out microbes. Poultry keepers can successfully simulate three of these requirements for incubation using an incubator. Keeping the incubator clean eliminates the need for the oil from the feathers.

Hatching times

Whether you are using an incubator or a broody hen to incubate eggs, they will take the same time to hatch – 20 days for Bantam eggs and 21 days for Large Fowl eggs. Some of the True Bantams have been known to hatch at around 19 days. If you set the eggs after 4pm, calculate the hatch date as if you had set them the following day.

Incubating eggs

Stage 1
Gather the required number of hatching eggs together. Do this over a maximum period of 11 days and store the eggs in a cool place during this period.

Stage 2
Set the eggs under a broody hen or in an incubator. Calculate hatching date from this time.

After 21 days of incubation the eggs should hatch.

7-11 days

21 days
(20 days for Bantams)

Candle the eggs at 5-7 days and at 12-14 days to check they are developing properly (see page 52).

Broody hens

Using broody hens is the natural way of incubating eggs. Some breeds of poultry are natural mothers and are more than delighted to go broody at every opportunity. Other breeds consider motherhood a very eccentric and strange activity to aspire to. Needless to say, you require one of the former types when wishing to hatch chicks.

This is where certain cross-breed birds are useful. The cross-bred Silkie is world-renowned for making the best broody mother. The pure-bred Silkie is also excellent, but chicks can sometimes become entangled in her feathers, so by out-crossing her with a normal feathered breed, the resultant female offspring will make safe and reliable broody hens. Silkies are often deliberately crossed with sitting breeds, such as Sussex, Wyandotte or Gamefowl. Light Breed hens are usually less reliable and may even abandon the eggs half way through the incubation period.

A hen will only go broody after she has laid some eggs; if you remove her eggs daily, the chances are that she will simply continue laying. To encourage a hen to go broody, try leaving some eggs in the nest box for her to sit on, or leave artificial eggs there.

Making a broody box

Do not allow a broody to try to hatch her eggs in a communal poultry house. There is a strong chance that the other birds in the run will ruin the clutch of eggs she is sitting on. She will need somewhere separate to carry out her brooding.

This need not be an elaborate affair; a dark, ventilated box that measures about 45cm^2 and is free of vermin will suffice. Many a brood of top-quality birds has been produced from a nest in an old tea-chest or cardboard box. As long as it is nice and comfy the hen will not mind. The sleeping compartment of a rabbit hutch serves the purpose well. The run of the hutch also provides an area to leave food and water for the broody as required.

Cut a turf to fit the base of the nest, turning it earth-side-up, then scoop out the centre to hollow it slightly so that the eggs will stay towards the middle of the nest. The turf will help provide the necessary humidity for the nest. On top of this you can place some soft, clean, dry and leafy-type hay. Use enough to make a soft bed that is slightly concave in the centre, but not so much that you risk the eggs becoming lost in it.

A broody box

The door should allow you to remove the hen without lifting her upwards.

Ventilation is essential for the comfort of the broody hen.

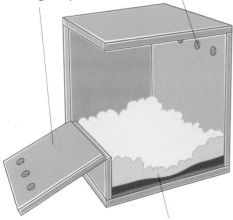

The hen will mould the nest into the shape she feels is best.

Pekins make excellent and caring broodies. They are True Bantams.

Below: *If a hen remains sitting in the nest box after laying, it usually indicates she is starting to go broody. This nest box has three compartments.*

Preparing broody hens

While waiting for your hens to go broody, take the opportunity to dust or spray the potential mothers for lice and mites and make sure they have been treated for scaly leg mite. Sitting on eggs whilst itching and being bitten to death is not any hen's idea of fun. Even good broodies will abandon their eggs if either they or the nest become infested with mites. And scaly leg mites will quickly infest the young chicks if they hatch under a broody that is infested.

Indications that a hen is starting to go broody will be evident when she starts to spend a longer time than normal in the nest box after laying. She will then become reluctant to leave the nest box altogether. She will fluff her feathers at you and make a soft growling sound, and may well peck at you if you try to move her. If you put a hand underneath her she may push downwards on it.

49

Sitting, hatching and feeding chicks

Move the hen to the broody box after dark. You need to be sure she is going to sit well for you, so rather than risk any precious eggs under her straightaway, place some artificial eggs or golf balls in the nest. If she is really broody she won't puzzle why she is sitting on golf balls, she will sit anyway. Shut the door so that she has to stay on the nest and leave her to sit quietly without disturbing her. Check her late the following afternoon. If she is still happily sitting you can lift her off the nest and place her in a run to stretch her wings, have some food and a drink and relieve herself, before reintroducing her to the nest. You will soon discover that she stores up all her excrement to release at this one time when she is let off. Not surprisingly, the smell from broody excrement can be rather overpowering at times! Her breaks should only last about 10 minutes a day, but towards the end of incubation this can increase slightly to about 20 or even 30 minutes.

The broody may be a bit reluctant to go back to the nest the first time you remove her as it is not where she herself has chosen to nest, but a good broody will quickly accept that her nest has simply moved location. After another day you can put your clutch of eggs under her. The best way to judge the correct number of eggs to put under a hen is to place them underneath her one at a time until she is just able to cover them all without having to stretch herself; then remove one egg.

Apart from making sure the broody comes off the nest once a day, you can now leave the rest to her. She will maintain the right temperature and humidity, and even turns the eggs. She instinctively knows far better than you do how to hatch an egg.

Occasionally, a broody hen will turn out to be unreliable and gives up brooding, but if you select your broody hen well, and use a proven type, she will generally be the most effective and environmentally friendly incubator you can find.

Hatching

When the chicks start to hatch, leave the broody undisturbed to get on with her job. All the chicks

Left: The sleeping compartment of a rabbit hutch makes a good place for a broody to site her nest. Here a Bantam Buff Orpington broody has raised two Bantam Lakenvelder chicks.

Above: *A Gold Silkie hen very busy with her clutch of youngsters. Silkies are gentle and docile chickens that are very good broodies and ideal for beginners.*

should have finished hatching within 24 hours. After this, hold any eggs that are left to your ear. If you can hear a tapping noise, replace the eggs and let the broody sit a while longer. You could also tip the eggs gently from side to side. If you can detect any flopping motion, it means the egg is addled and should be removed. Once you are sure the hatch is complete, discard the remaining eggs as they will not hatch.

Place the broody and her youngsters in a run or cage with dry wood shavings on the bottom for a few days. Keep this in a shed so that there is no chance of the youngsters getting wet. It is too soon for the broody and chicks to be outdoors; trailing along through wet grass will easily chill a chick and kill it.

Feeding and protecting the chicks

You will need a heavy dish to hold the chick starter crumbs. As you will soon discover, the broody will promptly start scratching all the contents out of it in her eagerness to feed her chicks. A very heavy

stoneware dish with a low rim can prove useful as it is difficult to tip over. Lightweight plastic dishes are easily turned upside down and can trap a chick underneath. Also provide a chick-sized water fount. (See page 56 for more information on feeding requirements for chicks.) The broody hen will teach them where to find the food and water.

After about a week you can move the broody and chicks outside onto very short grass. Be sure to protect them from predators by keeping them in a secure coop, as crows and magpies will readily swoop down and take a young chick before the broody has time to protect it. Domestic cats can also prove a problem at this stage. When the chicks are seven or eight weeks old, the broody decides that continual motherhood is a vastly overrated experience and tires of the chicks. At this point you can move them to their own accommodation.

Above: *A hen will call the chicks to her to show them the food. This Buff Sussex mother has the full attention of her chicks as they peck at chick crumbs.*

Artificial incubation

Using incubators is nothing new. The Ancient Egyptians had developed the practice by 1400 B.C., using specialist incubators to hatch many thousands of eggs at a time. Breeders retained a percentage of chicks that hatched as payment for the service. The Egyptian practice of egg incubation is documented in Roman literature, and in the first century B.C. Diodorus Siculus, a Greek historian, reported on the use of Egyptian incubators, stating "The men who have charge of poultry and geese, in addition to producing them in the natural way known to all mankind, raise them by their own hands, by virtue of a skill peculiar to them, in numbers beyond telling; for they do not use the birds for hatching the eggs, but, in effecting this themselves artificially by their own wit and skill in an astounding manner, they are not surpassed by the operations of nature."

However, the art of artificial incubation was not taken up by the Western World with any real enthusiasm until the beginning of the twentieth century. And it was later still that backyard poultry keepers really took the method to heart.

Buying an incubator

The most important advice you can ever be given about using an incubator is: follow the manufacturer's instructions!

Buy the best incubator you can afford. Each type has its plus and minus points. A forced air incubator, in which a fan keeps the air moving, will maintain a constant temperature but can be a bit drying when it comes to hatching. A still air incubator, which is better for hatching, can be rather more difficult to maintain at a reliable temperature. In any case, choosing a model with an automatic turning mechanism is well worth the initial investment.

Candling the eggs

When the eggs have been set for five to seven days you can candle them to see if they are fertile. Working in a dark room, hold each egg up to a bright light. If it is fertile you will be able to see a network of tiny veins starting to develop in the egg. An infertile egg will look exactly the same as a newly laid egg. These infertile eggs are called 'clears'. Remove any 'clear' eggs; leaving them in the incubator risks harbouring infections. Candle the eggs again at 12 -14 days and you will be able to determine the developing chick quite easily. Once again, remove any eggs that have obviously stopped developing or in which the embryo has died.

Left: *Starting to show embryo development.*

Right: *This egg shows no development.*

Using the incubator

It is vital to site the incubator in a room with a constant temperature, as fluctuating room temperatures can affect the overall temperature of the incubator and have a very bad effect on the hatchability of your eggs. Pre-warm the incubator and run it for about 24 hours to ensure it is operating correctly before setting any eggs in it.

Place the eggs in the incubator so that they rest on their sides or set them with the pointed end downwards, depending on the setting trays provided. If the incubator does not have an automatic turning mechanism, you will need to turn the eggs three times a day for the first 18 days of incubation. Do not always rotate them in the same direction as this will cause the chalazae to wind up like a spring and may reduce hatchability.

If you have chosen a still air incubator, set eggs that are of similar size. Putting eggs of different sizes, such as those of Large Fowl and Bantams, in the same tray can adversely affect the hatch rate. This is not so important in a forced air incubator. Always set the eggs so that they will all hatch at the same time. Still air incubators are run slightly hotter than forced air incubators to compensate for the layers of warm air that develop. Measure the temperature in these incubators just above the surface of the eggs.

All manufacturers will have tested their incubators rigorously and know how to get the best hatches from them. You will be told when to add water to the trays to increase humidity, and when to open or close the air vents to control air flow past the eggs. By strictly following their advice, and providing your eggs are fertile and the adult stock fit and healthy, your chicks should hatch successfully.

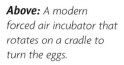

Above: A modern forced air incubator that rotates on a cradle to turn the eggs.

Left: The automatic turning mechanism in this fan-assisted incubator will turn the eggs every hour for the first 18 days. Then switch off the turner to allow the chicks to move into the correct position for hatching. Clean and disinfect your incubator between hatches.

Hatching and raising chicks

When your chicks start to hatch, whatever you do and no matter how tempted you are to help out or pick up a chick, DO NOT OPEN THAT INCUBATOR DOOR! In fact now is a good time to go shopping, visit a relative or friend or even go the cinema. Opening the incubator part way through the hatch is the surest way to kill off any late-hatching chicks.

The incubator will have built up the humid atmosphere required for the chicks to hatch. Opening it to handle the first chicks out of the shell allows all the humid air to escape and it will take up to half an hour for the humidity to build again. This is just long enough for the membrane around the as yet unhatched chicks to dry out and trap them inside the shell. Newly hatched chicks will be fine

Below: At the end of incubation, you can expect to find a number of eggs that have failed to hatch. These are referred to as being 'dead in shell'.

in the incubator for 24 hours. They will use the feed supply from the yolk that they absorbed to nurture them through this period.

What is wanted: a strong healthy chick, cheeping happily.

Raising the chicks

Once the hatch is complete, remove the chicks to a pre-warmed brooder area with a bed of wood shavings in it. A brooder is basically a box with a heat lamp or some other form of heating. Newly hatched chicks are unable to control their body temperatures at this age and becoming chilled can have fatal results. It can so weaken a chick that it may even succumb a few weeks later, so keeping your chicks warm is imperative.

The temperature in the brooder should remain close to 35°C for the first week. An infrared heat lamp or a device called an 'electric hen' will serve well for this purpose. Reduce the heat over the following weeks by about 3°C per week until the temperature reaches 20°C or so, by which time the chicks should be fully feathered. Do this by raising the lamp away from the chicks by a few centimetres each week.

The age at which the chicks can come off heat varies according to the time of year, the ambient temperature and the facilities you provide, but anywhere between five and eight weeks is normal.

You will quickly learn if the temperature is too high or too low by observing the behaviour of the chicks. If they try to get as far away as possible from

A hay-box brooder

It is possible to get a batch of chicks away from artificial heat at two or three weeks of age by using a hay-box brooder. This is basically an insulated box that uses the chicks' own body heat to generate enough warmth to prevent them chilling. This is only efficient for larger batches of chicks and should not be used for fewer than 15 chicks, as they would not generate sufficient heat.

If using a hay-box brooder, the time-consuming part is teaching the chicks where to go for their heat. This means careful monitoring for the first 48 hours or so. It involves letting them out of the brooder for food and water a number of times each day and then putting them back into the brooder compartment. Once they learn that it is warmer to go inside the brooder, they will do this when they feel the need for warmth, just as they would go under a heat lamp. Slowly remove the insulation from the brooder as the chicks grow. Although hay-box brooders are not widely used at present, increasing electricity costs, plus modern insulation materials and growing environmental awareness, can make them a very cost-effective and environmental way of raising chicks in the modern age.

A hay-box brooder

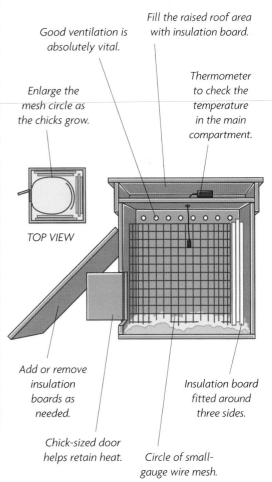

Good ventilation is absolutely vital.

Fill the raised roof area with insulation board.

Enlarge the mesh circle as the chicks grow.

Thermometer to check the temperature in the main compartment.

TOP VIEW

Add or remove insulation boards as needed.

Insulation board fitted around three sides.

Chick-sized door helps retain heat.

Circle of small-gauge wire mesh.

the lamp you need to raise it, as they are too hot. If they are huddled up close underneath it you should lower it, as they are too cold. When they are evenly spread out the heat is just right.

Listen to your chicks and you will soon recognise if they are happy or not. A high-pitched rapid and loud sound indicates that something is not right with them; maybe they are hungry or need water. A low-toned 'chatty' cheep means they are contented. A chick accidentally separated from the rest will not be shy to announce its feelings.

Feeding chicks

Feed chicks on a proprietary brand of chick starter ration or chick crumb until they are eight weeks old. This food is in the form of crushed pellets and is easy for them to eat. It usually sold in a medicated form to protect the chicks from a protozoan disease called Coccidiosis (see pages 71-72). The low level of medication in the feed does not prevent them catching Coccidiosis, but usually keeps any symptoms at bay until the chicks have built up their own immunity. It is possible to avoid using the medicated feed, but you need to be extra rigorous with litter removal and management.

Use chick-sized water founts rather than any other type of drinker. They are specially designed to allow the chicks to drink without falling in and drowning. A few of the incubator-hatched chicks should have their beaks dipped once into the water so that they discover where it is. The rest of the

Below: A group of healthy Orpington chicks in various unusual colours, including Spangled Jubilee and Lavender. These chicks are five days old.

chicks will then copy their example. Do not be over-zealous in this practice; the chicks should not ingest a lungful of water. A broody hen will also be quite content to use a chick-sized water fount.

If the chicks are raised with a broody hen, it is perfectly natural for her to eat their food. It will not harm her at all. In any case, if you try to give her a larger type food she will only pick it up and offer it to the chicks. A treat such as a little chopped hard-boiled egg is a very good food for young chicks, as is a little chopped chickweed, harvested from the garden.

From the age of six weeks offer the chicks flint grit. This is especially important if you intend to add grain to the diet. Ordinary poultry flint grit is suitable for the chicks of Large Fowl, as they will pick out the smaller pieces, but you may wish to use pigeon-sized grit for Bantams.

From eight weeks old, the chicks are referred to as 'growers'. Now is the time to change their food

to a lower-protein pellet called a 'growers ration'. To avoid causing digestive problems at this stage, introduce the chicks to their new diet gradually over a few days, as they will need to get used to the different-sized food. Try adding some growers pellets into their chick crumbs for a couple of days before switching them over completely. They can remain on this ration until they move on to layers ration at 16-18 weeks old.

If you intend to feed grain, introduce this to the diet very gradually and in limited quantities over the space of a week or so after the youngsters have moved over to their new growers diet. Do make sure that grit has been freely available for a while before introducing any type of grain ration.

Foods for chicks

Chicks always appreciate chopped hard-boiled egg.

Feed chick crumbs for the first eight weeks.

The slightly larger pellets of growers ration.

Flint grit is important for the digestion of grain.

Below: These chicks are being brooded under an infrared heat lamp. Raise this higher each week to reduce the heat provided. The hanging feeder behind the water fount is a small version suitable for chicks.

Sexing chicks

In its first year, a male chicken is called a cockerel and a female is called a pullet. After a year old they are referred to as cocks and hens. Be aware that some breeds are far easier to sex than others and some are notoriously difficult. Certain breeds that are easy to sex at three weeks of age can prove more difficult to sex at five weeks, when they go through a different phase of growth. Here is some basic guidance to sexing chicks.

• Always compare like with like. For example, it is no use comparing a Sussex with a Minorca as the two breeds mature at different speeds.

• Cockerels start to develop their combs and wattles sooner than the females and usually have thicker legs. If you examine the developing saddle

Above: *The larger comb and wattles on the Lavender Pekin chick on the right easily mark it out as being a cockerel. These chicks are six weeks old.*

feathers near the tail and over the back, you will see that they are pointed in shape and not rounded. These feathers also often appear glossier than those of the pullets. Males sometimes have a slightly more petulant sound to their voice.

• Males with the Partridge pattern often show a glimpse of their adult feather coloration along the back and breast at a fairly young age. Males are often bolder, and more likely to play fight at an early age. (Pullets may also do this but less often.)

• Cockerels usually start to crow at 10 to 12 weeks old, but this varies with the breed. There is no hard and fast rule for the age at which this behaviour develops; cockerels have been known to crow as young as seven days.

• Pullets have a less developed comb and wattles and more delicate legs. The saddle feathers on the back, close to the base of the tail are rounder than in males. In breeds with a crest, such as the Silkie, the crest is often more rounded at an early age and will not show the longer pointed feathers seen in the males' crests. A pullet's voice is often a little lighter in tone than the male's.

Of course, it is possible to end up with an entire batch of one sex. If this happens you will have to wait until the birds are at the stage of developing secondary characteristics, such as the sickle feather shape, rather than rely on comparing individuals against one another.

Confident sexing of chicks is an art that takes time to learn, but is a very valuable one once mastered.

Some breeds carry a gene that helps to determine their sex by the rate of feather growth. In this case, the cockerels grow their feathers far more slowly than the pullets, remaining scantily clothed in their baby fluff, whilst the pullets develop wing and tail feathers long before their brothers do.

Left: *The white colour in the Light Sussex is caused by the silver gene. A hen (shown here) crossed with a cockerel carrying the gold gene produces sexable chicks.*

Autosexing breeds

A number of specially developed pure breeds have been created called autosexing breeds. In these, the genes used in the breed makeup result in the chicks being different colours at hatching. Understandably, this is very useful if you do not want to become fond of a chick only to discover a few weeks later that it is a male.

A similar effect can be created by outcrossing breeds carrying the silver gene and the gold gene – hence the popularity of the Rhode Island Red (gold gene) and the Light Sussex (silver gene) cross for producing commercial breeds. By using a Rhode Island Red male and Light Sussex females, all the buff-coloured chicks will be females and the yellow ones will be males. However, crossing the adults the other way around does not produce the same sex-linked effect and all the chicks will be yellow.

Below: *The Rhode Island Red (here a cockerel) carries the gold gene.*

Out-crossing to produce sexable chicks

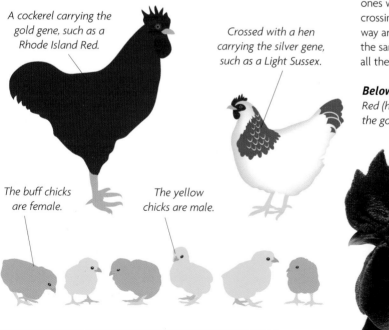

A cockerel carrying the gold gene, such as a Rhode Island Red.

Crossed with a hen carrying the silver gene, such as a Light Sussex.

The buff chicks are female.

The yellow chicks are male.

Managing growers

Never overcrowd growing stock, as this can lead to vices such as feather pecking and even cannibalism. The youngsters' newly sprouting feathers provide a rich source of blood, which can prove irresistible to other fowl in their pen. Providing youngsters with other interesting things to peck at, such as some

Below: *The Dutch hen below will soon tire of her little family, as they are old enough to fend for themselves during a trip around the garden lawn. She will then resume laying.*

old CDs hanging in their pen, will help to keep them occupied. If at any time you find a grower has been pecked, you must remove it from the pen at once to a place of safety, as the others will continue to peck at it with potentially fatal results.

Check the rest of the stock and if you find any others that appear to have been pecked, remove them too. Treat the pecked area with a product such as Stockholm tar or spray it with an anti-pecking spray. Once the pecked grower has recovered, coat the target areas with Stockholm tar before returning

the bird to the pen. Watch carefully for further signs of aggression. Once out on free range, birds rarely indulge in feather pecking.

Separate young cockerels from the pullets and house them in different accommodation at about 10 weeks of age, otherwise their increasing testosterone levels will cause them to pester and bully the pullets.

Moving the growers outdoors

When moving growers outside onto grass for the first time, be aware that, tempted by the novelty of this new green food, they may try to gorge themselves. Too much grass at once can impact

Above: *These two Black Wyandotte Bantam chicks are still young enough to require the attention and warmth of their Large Fowl Marans broody hen.*

the gizzard, resulting in death. So before letting the growers out of the house onto grass for the first time, make sure they have filled their crops with their normal pellets and that the grass is short. If you have recently cut the grass, rake up the clippings from the pen and only introduce the youngsters for a few hours on the first day. Once the novelty of eating grass has worn off, they will only peck at what they require, and being able to run outside will be the healthiest way to raise them to adulthood.

Understanding chicken behaviour

If you learn why your chickens behave as they do, it can help you understand their basic needs and cater for their environmental requirements..

Pecking order

Chickens must work out their pecking order for themselves. Basically, the dominant hen in the group can peck any individual, as they are all below her; the second in line can peck any chicken below but not the one above her and so on down the chain until you reach the bird at the bottom of the pecking order, who can be pecked by all the others.

The system works, and once a flock has established its pecking order it coexists peacefully. Introducing a new bird can upset this order, resulting in bullying of the new hen. Adding more than one hen at a time can ease matters, but put them in an enclosure first, so the others can meet them through the safety of some wire. It is always best to allow the new hens to meet in a place with

plenty of space, where they can run and escape the bullies until they become accepted. Providing hiding places and extra perches around the run can also be beneficial. You could also try removing the dominant bird and placing her in a new environment where the newcomers are already established before returning them all to the main flock together.

Pecking order

The dominant hen can peck all beneath her.

This hen can only peck one bird.

Left: *New birds often seek refuge from a bully by flying up onto a perch. Always allow plenty of perching space. You can make extra perches from tree branches of a suitable diameter for the birds you are keeping.*

Introducing chickens

Allow new hens to meet the flock through wire mesh at first.

The established flock will resent newcomers. New birds will have to find out where they will fit into the pecking order.

Provide some places for a new hen to hide.

It may take a while for the new hens to be fully accepted. Providing plenty of space in the run will help in the early stages of the introduction process.

However, beware of removing the lowest in the pecking order and putting her in a pen with smaller birds in the hopes of building up her self-esteem. She is very likely to turn into a total tyrant just to prove her new status.

Vocalisation

All breeds crow, but some smaller breeds are less noisy than others and in general the larger breeds tend to have a pleasanter-sounding crow. Cocks crow not only at dawn, but also throughout the day. It is most likely they are crowing for the same reason that songbirds sing at the dawn chorus. Not, as we may like to believe, for the joy of the song, but to see if any neighbouring bird has failed to survive the night and might have left a territory available for habitation. Cocks will also crow in defiance, in challenge and also to announce the

Crowing hens

Although reports of a crowing hen are usually greeted as newsworthy, this phenomenon is not unique. An item in the Edinburgh New Philosophical Journal in 1826 describes in great detail what appears to be a 12-year-old French breed hen that took on the demeanour, plumage and attitudes of a cock bird. The reason for the change is due to an ovary suffering damage of some kind, resulting in hormonal changes. Hens have sometimes moulted back into female plumage and resumed laying, but this is unusual; mostly the condition continues until the bird dies. Just enjoy the novelty of owning something rather different from the norm.

Inset: *All cockerels will crow, but some breeds are a lot more persistent than others. This is a Dandarawi cockerel about 12-14 weeks old.*

Left: *Siting the house within an area of thick vegetation helps dull the sound. Here, a dense planting of trees and shrubs will help to suppress noise.*

'all clear' signal when danger has passed. They will respond to the crowing of another cock with a blast of their own.

Cocks generally flap their wings just before crowing, which is all part of announcing to the world that they are fit, vigorous males in prime condition.

You can alleviate the noise nuisance to your neighbours by removing any offending birds to a dark box at night and placing them somewhere confined and soundproof, returning them to the run in the morning at a reasonable hour. Planting conifers around the run can help dull the sound, as will planting bamboo that provides a background rustling noise and makes the crow appear less noisy, even if in reality it is not. A group of the lower-toned wind chimes can have a similar effect. Chickens have many different sounds and it is worthwhile and informative to learn to recognise them. Hens issue a soft growling sound when broody, cackle to announce when they have laid an egg and make a 'took-took-took' noise to call their

chicks for food. The cock makes a similar noise to call the hens for food. Chickens will screech when alarmed and make a burring sound to alert others to overhead dangers. Listen to your birds and you will find that they really do say quite a bit.

Courtship behaviour

A cock bird will call the hens to him if he finds an interesting treat. The hen is not stupid; she knows full well it is not just food that is on the male's mind. Although some males will pay gentlemanly attention to the hens all year around, the chances are that the male is just using the food as a bribe to lure the hen within mating range.

If the hen accepts the food, the cock then waltzes around the hen with one wing dropped before attempting to mate. If the hen is in lay or approaching lay and prepared to accept his advances, she will squat to allow mating to take place. Hens that are not in lay are quite likely to accept the gift of the food and then make a hasty retreat before they can be caught.

Aggression

Some males become very aggressive during the breeding season in spring and summer. Although this may appear amusing to an adult – especially if the bird in question is a tiny Bantam that is batting furiously at their feet – for a child it can be terrifying. If the bird aims its attack at the head and eyes it can also be dangerous. Unfortunately, the most aggressive birds are often the most virile and fertile, but unless needed for breeding, Large Fowl that are overly aggressive are best consigned to the oven. They hit too hard and jump too high, and can easily cause injury. Bantams are easier to manage but can still cause problems.

Many people recommend ways of curing the habits of antisocial birds. Some suggest throwing water over them or dipping them in a bucket of water. Others advise cuddling them, which is great – providing their beak is not latched onto a vulnerable part of your anatomy at the time. Although many methods work as a temporary measure, the birds are soon back to their 'beat up the human' game and the reason is simple. You have to understand that because you eventually walk away after an attack the bird always believes it won that particular fight.

When you need to catch an aggressive male, hold a coat in front of you for protection and throw it over the bird as it approaches. Then pin the bird to the ground inside the coat.

If you want to keep the bird for breeding, house it in an enclosure that you do not need to enter often, perhaps a pen that you can tend from the outside. Otherwise, you will just have to put up with any damage to yourself. With luck, the male may well calm down again outside the breeding season.

Lastly, do not think of passing an aggressive male on to an innocent buyer without warning the prospective owner of the bird's temperament.

Above: A small but feisty cockerel squaring up to peck. After an attack the human eventually walks away, making the bird think it has won the fight.

Above: And so it will attack again the next time you arrive. After you leave, the bird will almost certainly start to crow, announcing his 'victory' to the world.

Egg eating

If you collect your eggs on a regular basis you are less likely to encounter the problem of egg eating, but when it does occur it can be infuriating. No sooner has the hen or one of her companions laid an egg, than the culprit will pounce, peck open the egg and promptly eat the contents.

Egg eating is readily copied by other hens and you must intervene rapidly, otherwise you will soon find the entire flock standing around the nest boxes waiting for the next tasty snack to arrive. Never allow your hens to find out that eggs are good to eat!

If you can identify the hen that is doing the egg eating, remove her from the pen at the earliest opportunity, before she teaches her bad habits to any other birds. If you are not sure which hen is the culprit there are other options available. Although the old trick of filling an egg with mustard may help, chickens actually have a very poor sense of taste. The following tactics can prove more successful.

Ways to prevent egg eating

Firstly, make sure that the entrance to the nest box is in the darkest part of the hen house and facing away from any window, thereby making it more difficult for the hen to see the eggs. You can also make the nest boxes darker by hanging a fringe of material over the entrance to the nest box.

Secondly, make sure that the wood shavings in the nest boxes are a good 7.5-10cm in depth. Fluffing up the shavings a bit allows the eggs to sink down

Below: The white egg here is a fake one and can help to curb egg eating. Siting the nest box in the darkest part of the house is also a good strategy.

A plastic roll-away liner can be fitted to a nest box.

The slope means the eggs roll to one end.

Above: *When using a roll-away liner such as this, make sure the egg collection area is covered to prevent the hens accessing the eggs.*

Shading the nest box

A fringe of material will help make the nest box dark.

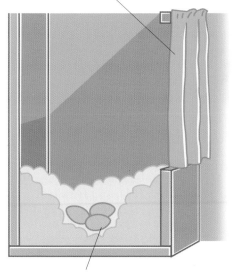

The eggs will sink into the soft bedding and be lost from sight.

into them and become less visible. If your birds lay dark brown eggs, mix some dry peat with the shavings to help hide the eggs. Place a quantity of fluffed-up hay on top of the shavings.

Finally, leave a few fake eggs around the chicken run. Choose solid, quite heavy ones, made from either rubber or china, which are often available from agricultural merchants. Hens are not usually fooled by golf balls, but you could try using them if you have problems finding artificial eggs. Initially, the egg-eating hens will peck at the fake eggs as they try to open them up, but eventually become bored with the idea and give up. Hopefully, they will also leave the real eggs alone.

If your hens continue to egg eat, you could consider installing roll-away nest box liners in your nest boxes. These devices are available commercially at a reasonable cost and ensure that as soon as the egg is laid, it gently rolls away to a place of safety for you to collect later.

Health care

It is important to state from the outset that the information offered here is not designed to replace the advice of your veterinarian, but is intended to give you an idea of what may be wrong with your bird. Only the most basic ailments are described.

If you suspect a problem, DON'T PANIC. Just like anyone delving through a medical dictionary, it is easy to come quickly to the conclusion that your chicken is suffering from every ailment under the sun. The chances are that it has one of the more common complaints, rather than something strange and exotic.

If a bird looks ill, the first thing to do is separate it from the others. Chickens are no respecter of ill-health and will happily take advantage of a sick bird if the chance arises by attacking and injuring it. Place it somewhere quiet where it can rest. The second thing to do is provide warmth. A heat-lamp or heat pad for any ill birds is a wonder drug in itself. A bird that is off-colour can often revive after a day of sitting under a heat lamp.

Worms

Poultry suffer from several types of worms, with the roundworm being the most common. Worms in poultry can cause various problems, including diarrhoea, anaemia, weight loss and poor egg production. Chickens with a heavy worm burden generally look rather unwell, with a ruffled and droopy appearance. Birds kept on free range rarely suffer from such a large worm burden that it actually causes major health problems. However, worming your birds twice a year should ensure that they stay healthy and will help reduce the overall worm burden on your land. Birds kept more intensively may need more frequent worming.

If you intend to show your birds, do not treat them for worms during the moult, as wormers may prove detrimental to new feather growth. Wormers are available from your veterinarian or from licensed agricultural suppliers.

Lice

Lice are small chewing insects that feed on feather scales, dry skin, debris and scabs. They often gather around the bird's vent, although they can be found anywhere on the body. They are fast moving, light brown in colour and flattish in shape. They leave their egg clusters glued onto the feather shafts. Lice are exceedingly irritating and will affect the overall condition of your birds. You may very well find scabs where the birds have been frantically scratching themselves. Luckily, lice are easy to kill by spraying or powdering the bird with one of the many products available for this purpose. Trim off any egg clusters on the feathers and burn them.

Mites

Several types of mite infest poultry; the following three are particularly troublesome: the scaly leg mite, the northern fowl mite and the red mite.

The scaly leg mite burrows under the scales of the legs and causes unsightly raised scales. This can cause the bird pain and if the infestation becomes severe it can even cripple a bird. This mite is best treated either with Ivermectin, obtained on prescription from your veterinarian, or by smothering the mites with a substance such as petroleum jelly, which stops the mites breathing.

In severe cases a better recovery can be achieved by applying the petroleum jelly, wrapping some tissue paper around the leg and then bandaging over this with waterproof tape. Leave the bandage in place for a few days. This helps keep the petroleum jelly on the leg and kills the mite more efficiently. It also helps soften the hardened growths so that they come away more easily.

Northern fowl mite appear as a seething mass of pinhead-sized moving bodies, usually around

Below: Applying a lice powder treatment. As well as powdering around the vent area, treat the body as well.

Right: Lice eggs will remain glued to the feather shafts unless removed manually. Then burn the egg clusters.

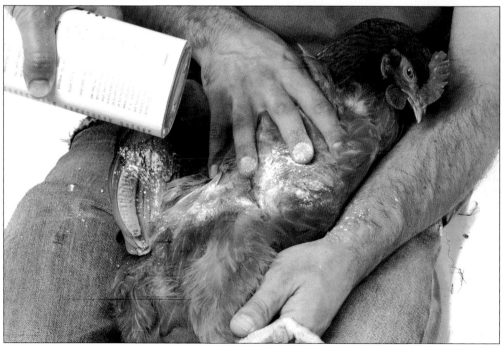

the vent area. They feed on blood, causing severe loss of condition, and can transmit other diseases. Birds suffering from northern fowl mite often have blackened dirty feathers and scabs around the vent area. These mite also inhabit the crests and beards of breeds so adorned and they can prove troublesome to eradicate. Your veterinarian may be prepared to prescribe an Ivermectin-based product for you to use. Although this product is not licensed for general poultry use, veterinarians will sometimes prescribe it and it does prove effective.

If you use a traditional powder or spray, make sure you also dab some around the bird's ears, being careful to avoid the eyes. These mite can sometimes take refuge by hiding in the ear canal and evading the usual topical treatments, thus permitting

Left: *This is a mild case of scaly leg that can be easily treated. If left, it would get far worse, causing the bird pain and even crippling it in time. Examine your birds on a regular basis.*

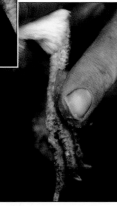

Right: *If you are treating a bird for scaly leg using petroleum jelly, make sure the entire leg is well covered.*

another generation to take hold a few weeks later. Oddly enough, if your bird has lice it tends not to have northern fowl mite. It seems these two parasites dislike the presence of one another and will not live on the same host at the same time.

Red mite are similar to northern fowl mite, but spend most of their life-cycle living in the cracks and crevices of the hen house, emerging at night to feed off the blood of your chickens. They are normally grey in colour, only becoming red after a meal of blood.

If your hens suddenly take to roosting outside on the hen house roof, refusing to go indoors at night, or emerge from their house in the morning with pale faces, it is almost certain that the hen house has become infested with red mite. To check for this, after dark rest your hand on the perch; then remove it and study the horrid, scurrying little grey mites that are rushing all over your hand. Alternatively, look for blood spots on the eggs or wipe a paper tissue along the perch and see if it comes away with streaks of blood on it.

Since the banning of creosote for domestic use, red mite has become a truly major pest of backyard poultry. New sprays and potions appear on the market almost daily, most which have a short-term knock-down effect; others claim to have a longer residual effect.

A severe red mite infestation can drain a hen of enough blood to kill it. You will frequently find the mites loitering in large numbers under and around the ends of the perches. Look for a white powdery substance – the mite's droppings – as a tell-tale sign of their presence.

In spite of their almost mythical reputation for surviving chemical deterrents, red mite are not immune to physical attacks. A pressurised steam cleaning of the hen house can have as much

Left: Red mite are the true vampires of the poultry yard. They will take up residence anywhere they feel they will not be disturbed. Treat any red mite infestation without delay.

effect as many chemical sprays and is far more environmentally friendly. A gas blow torch directed at the corners and crevices of the house is a most satisfying retribution, but do be careful not to burn the house down!

As all TV and film vampire hunters will tell you, the best time to find vampires is after dark. Therefore treat the henhouse at night, when the mites have emerged from hiding in the crevices. (A good tip is to wear a shower-cap to stop them getting in your hair.) In the meantime, put your birds in a cardboard box for the night somewhere safe. In the morning, treat the birds with an insecticidal powder and burn the cardboard box, as it will contain any mites that have dropped off the birds during the night.

Repeat the hen house treatment the next night, and then every third day for a minimum of two weeks. Using food-grade diatomaceous earth in the house and on the birds can help to prevent reinfestation. See if your veterinarian will also prescribe an Ivermectin drop-on treatment for your birds.

Many people report that red mite appear to dislike the smell of phenol-based disinfectant, and disinfecting the house when cleaning can have a

good repellent effect. The total life cycle of a red mite can be as little as a week and they can quickly multiply into millions. By persistently killing the emerging larvae over the course of two weeks you can eventually break the cycle.

Using a combined physical, chemical and a repellent approach appears to be the best way of dealing with these horrid beasties.

Coccidiosis

Coccidiosis is a protozoan disease caused by a single-cell parasite that lives in the wall of the chicken's gut. There are seven different types that affect chickens. It is mostly found in young chicks around three to six weeks old, but it can also affect older birds. It is passed on via the droppings and can continue to be infective for around a year.

Coccidiosis thrives in warm damp conditions, so by keeping the chicks' bedding dry and changing it frequently you help protect against it. Feeding medicated chick crumbs will usually keep it at bay until the chicks have developed natural immunity.

Chicks suffering from coccidiosis usually look very miserable and hunched, with ruffled feathers, and often cheep pathetically. One well-known symptom is the passing of blood in the droppings, but by the time this stage is reached, you can expect a few casualties and not all types of coccidiosis produce this symptom.

Pay close attention to the youngsters in warm humid weather, as this is often when the disease strikes. An outbreak may well occur if chicks are placed outdoors onto previously contaminated ground before they have developed immunity.

If you suspect your chicks may have coccidiosis then early treatment with the correct drug is absolutely essential. Your veterinarian will prescribe the necessary drugs to treat an outbreak and these take effect fairly quickly. Even if the chicks were off-heat at the time of the outbreak, reintroduce a heat lamp, as this will aid their recovery.

It is possible to vaccinate against coccidiosis, but this can prove expensive for a small batch of chicks.

Respiratory infections

Birds do not suffer from hay fever, so if you hear sneezing, wheezing or coughing or if your birds have snotty noses or make a rattling noise when

Below: *The swollen sinus (just below the eye) is a clear sign that this bird has a respiratory infection.*

they breathe, it means they have picked up either a viral or bacterial infection or both. Such infections are quite common and do not mean your birds have avian influenza. One of the most common infections of poultry is a bacterial one called Mycoplasma. This can appear by itself or alongside other infections and cause the symptoms to become worse. Whatever infection your birds have contracted, they will need treatment. Thankfully, Mycoplasma infections react well to treatment with antibiotics, so if your birds are coughing, sneezing or wheezing, see your veterinarian.

Egg binding

An egg-bound hen will show signs of distress. She will usually be standing like a penguin and be visibly straining. Egg binding is usually caused by an egg that is rather too large or malformed for the hen to lay easily. Heat is excellent for helping egg-bound hens. Often, simply by placing the hen somewhere quiet for a couple of hours and providing her with a heat lamp or a heat pad she will manage to lay the egg. If it is badly stuck, then some manual assistance may be needed. If the egg is visible, try introducing some olive oil or petroleum jelly to the entrance of the vent to help lubricate the passage of the egg. Failing this, your veterinarian should be able to assist, as there are various calcium-based injections that can rapidly enable an egg-bound hen to lay the troublesome egg.

Crop impaction

An impacted crop can be caused by the birds eating too much long grass or rubbish, such as freshly mown grass clippings. Some people recommend treating the condition by flushing and emptying the crop via the beak. However, be warned that many birds are accidentally killed by people trying to flush out an impacted crop and allowing fluid to enter

Chicken digestive system

As a chicken is unable to chew its food, it relies on softening the food in its crop and then passing into a muscular grinding organ called the gizzard.

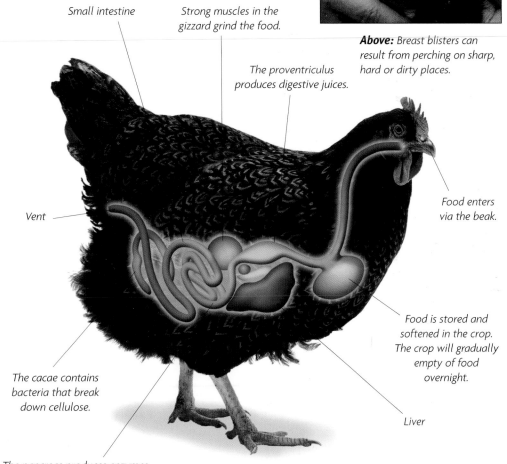

Small intestine

Strong muscles in the gizzard grind the food.

The proventriculus produces digestive juices.

Above: *Breast blisters can result from perching on sharp, hard or dirty places.*

Food enters via the beak.

Vent

Food is stored and softened in the crop. The crop will gradually empty of food overnight.

The cacae contains bacteria that break down cellulose.

Liver

The pancreas produces enzymes that digest fat and protein.

Below: The bulging crop seen on this hen should slowly empty overnight.

Crop

formula can help keep the bird nourished. You may find that given sufficient liquid nutrition and plenty of massage, it may recover on its own when the crop contents eventually break down. Otherwise, the only alternative is a trip to the veterinarian so that the impacted mass can be removed surgically. The prognosis for an impacted crop is very variable depending on how bad it is, and whether there is a gizzard impaction as well, which is far more serious.

Sour crop

Sour crop happens when there is a delayed emptying of the crop. The 'sour' part is generally caused by a fungal infection that takes hold because of this delay, causing a vile-smelling liquid to build up. If you tip a bird with sour crop forward, it will usually expel a great deal of this fluid, alleviating some of the pressure. Treatment should be with oral anti-fungal drugs from your veterinarian. Offering natural live yogurt can help the condition, but you should always look for the cause of the delayed emptying of the crop, as well as treating the sour crop itself.

Toe balls

Chicks and adults sometimes suffer as a result of balls of excrement or mud that cling to their toes. With chicks it is often caused by them stepping in

the windpipe. Be very careful if you decide to try this course of action.

If you suspect that your bird has an impacted crop, check it in the morning after a night of fasting. At this time the crop should be empty. If it still has a solid mass in it, then this suggests your bird may be crop-bound.

Try giving some olive oil and gently massaging the crop to help break up the mass. Provide liquid-based nutrition rather than hard food, as this may get past the blockage. High-calorie liquid or paste supplements used for dogs and cats, or baby parrot

Left: A bird with overly long toenails can be prone to picking up balls of dirt and excrement. These will need to be removed.

their water and food, which then hardens into solid balls encasing the toenail. If left, these can build up in the same way a snowball grows and can result in the loss of a toenail. Never pull off these toe balls, as the toenail may come with it. Soak the toes to soften the ball and remove it a small bit at a time. This can be a slow and delicate procedure with tiny chicks, but patience is the watchword here.

Above: *Be very careful when removing toe balls from chicks, as it is easy to damage or even pull off their tiny toenails.*

Pasted vents

In the case of young chicks, excrement can become stuck around the vent area so that they are unable to excrete any more faeces. Understandably, this can have fatal results. The problem is often the result of the chicks receiving a chill during the first 48 hours of life. Bathe and clean the mass with tepid water until it is removed. Then, very gently, try squeezing the vent to see if you can remove any backlog of faeces that may be waiting to be expelled. Once this has been done, dry the area and apply a little petroleum jelly to help stop further pasting.

Right: *The dirty feathers around this bird's vent suggest that it may well be suffering from mites.*

Showing birds

Just as dog shows are dedicated to the exhibition of pedigree dogs, poultry shows are dedicated to the exhibition of pure-bred poultry. People who show poultry are usually referred to as 'fanciers'. In her book The Poultry Yard, written around 1870, Elizabeth Watts writes of poultry fanciers: "Generally speaking the keeping of poultry was regarded as 'a useless hobby' 'a mere individual caprice,' 'an idle whim from which no good result could by possibility accrue;' nay, some times it was hinted 'What a pity they have not something better to employ them during leisure hours!' and they were styled 'enthusiasts.' "

All this is still true and the 'enthusiasts', one of which you may become, are still with us to this day.

National poultry clubs publish a standards book for all the recognised breeds, with a detailed written standard of points to which the breeds should conform. Most breeds also have their own breed club to represent their interests. Minority breeds without a club are collectively catered for by the Rare Poultry Society. These societies can be a source of great help and information. Find the standard for your breed and study it to make sure your birds meet it fairly closely.

Choosing a breed to show

When choosing a breed to show, it is worth bearing in mind the extra complications present in a few distinct breeds, where exhibition breeders practice what is called double mating. Basically this means keeping two different-looking strains of the same breed. One side of the strain will produce show-quality cockerels (the cockerel strain) and the other side of the strain will produce show-quality pullets (the pullet strain).

For example, you might consider breeding top-quality Minorcas for show. The standards call for the male to have an upright comb and the female a folded comb. Thus, the pullet breeding strain

---

Poultry rings

Just as many other breeds of livestock can have their pedigree or ownership traced, The Poultry Club of Great Britain runs a ringing scheme for pure-bred poultry. You can buy numbered rings (right) in the required size for the breed of poultry concerned. These rings help you keep a record of your breeding stock and if a bird is subsequently sold, it can be transferred to the new owner through the Poultry Club's record system. Rings should be fitted when the birds are quite young, although this varies from breed to breed. They will normally be fitted when the bird is around 9-10 weeks old by sliding the ring over the foot. It is not compulsory to ring show birds, but many breeders do so, as a rung bird is easy to identify. This can help if there are any doubts as to the bird's ownership.

Above: *Poultry will compete against others of the same breed in a show, such as in this class of White Leghorns in the top row of show pens.*

would be headed by a male with a folded comb and contain exhibition standard hens, also with folded combs. Although this breeding male would not be any good for exhibition himself, all his offspring would have folded combs, so this strain will produce females suitable for showing.

The cockerel breeding strain would be headed by an exhibition standard cock with an upright comb and contain hens also with upright (not folded) combs. The offspring produced from these parents will have upright combs, and thus this strain will produce cockerels suitable for showing.

Therefore, all novices should be a bit wary of purchasing a perfect show-quality trio of a breed that is usually double mated, as a pullet breeding strain crossed with a cockerel breeding strain will not reproduce the same top-quality show birds that you will find at the exhibitions.

If all this sounds far too complicated, then choose one of the many other breeds or colours that will produce show-quality males and females from just one breeding pen.

Entering birds for a show

Several weeks before the show, obtain the schedule from the show secretary and see what class your birds should be entered in. Certain abbreviations may appear in the schedule such as AOV (Any Other Variety), AC (Any Colour), AOC (Any Other Colour), LF (Large Fowl).

Be sure to check the closing date for entries; this will usually be a week or two before the show, but with the large shows it may be as much as eight weeks in advance. Make a note of the time the birds should be 'penned'. This is the time your birds should be ready and awaiting the judges attentions.

Naturally, all birds taken to shows should be fit, healthy, free from any types of parasites and in good feather condition. To this end, be sure to check them over thoroughly, looking for mites, lice, scaly leg, etc. or signs of illness.

Pen training

Any bird that is going to be shown needs to be well handled beforehand. Judges do not enjoy trying to remove an untamed bird from the show pen. It is worth noting that a bird that savages the judge is invariably marked down, especially if it affords any regular exhibitors a chance to have a good laugh at the judge's expense.

Therefore, you will need to embark on a period of 'pen training'. This means getting the bird used to being in a show cage. A suitably sized dog crate can be used for this purpose. Keep the bird caged for about a week, removing it often to handle and examine it in same the way as the judge. In addition to its usual food, provide some treats to encourage it to look forward to a hand being placed into the cage.

A judge will use a judging stick to turn the bird around and examine it through the bars of the show cage. When the stick is used the bird should react calmly, not take flight as if the judge were attempting to murder it. Again, this is something birds need to learn. Some breeds, such as Modern Game, are usually taught to pose when a judging stick is placed in the cage. Place a titbit on the end of a judging stick and a bird will soon happily accept having it introduced into the cage.

Preparing birds for a show

Birds should be washed about a week before a show. This allows the natural oils to get back into the feathers.

A judging stick can be used to move the bird into different positions.

Above: *When first being trained for show, a bird will often crouch. However, it will soon gain confidence to stand upright again.*

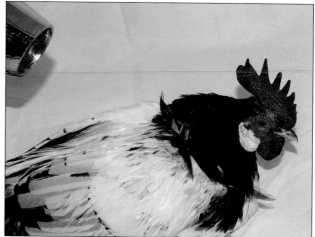

Below: Some birds will become very used to being washed for shows and will even start to enjoy the drying process. Use the dryer at a low heat setting.

Above: When washing a bird, make sure you do not get shampoo in the eyes. This means taking special care when washing the neck hackle.

Hard-feather breeds are not washed as often as soft-feather breeds, as it can soften the feathers, but they will still need to be presented in a clean condition. This is all part of showmanship; a dirty bird will not win prizes. Many are the times when a well-prepared inferior bird will go on to beat a poorly prepared but better specimen of the breed.

Clean the bird's legs with a soft toothbrush and use a toothpick or orange stick to clean carefully under the leg scales. Wash and sponge the head and comb of the bird to remove any dirt and debris, and then set about washing the bird itself.

For preference, use a cat shampoo or a baby shampoo that will not sting if it gets into the bird's eyes. The water should only be hand hot. Take care that the bird neither puts its head under the water nor tries to drink the soapy water. Thoroughly rinse the suds from the bird using slightly cooler water.

Carefully run your thumb and fingers down the wet feathers to remove excess water and then wrap the bird in a towel. Use slow, gentle heat to dry the bird. Most fanciers use a hair dryer or fan heater on a gentle heat setting. This drying will take much longer than you would expect, as the moisture can take a long time to come out of the underfluff of a bird. If at any time the bird starts to turn a bit purple in the face, turn off the hair dryer and rest for a while, as you are probably overheating it. Splash

Left: Feeding grain for two days before a show will keep the droppings firm.

79

Right: *Judges will have many birds to examine during the course of a show. An assistant keeps notes as the judge moves from pen to pen assessing the entries.*

Below: *The show judge examines each bird thoroughly to ensure it meets the requirements for the breed. This is a Japanese cockerel.*

a little cold water over the bird's head to refresh it. Never leave a bird unattended in front of a fan heater as it can easily overheat.

Once the bird is totally dry, place it back in the training pen on a clean bed of shavings. You may wish to feed fluffy breeds with wheat or mixed corn for the two days preceding a show, as this will make their droppings firmer and less likely to soil the feathers around the vent.

On the morning of the show, get up early to check over your bird before setting off. It is amazing how some birds manage to soil their feathers at the very last moment. It is far easier to do a quick fix at home, rather than in a panic at the showground.

At the show

Take your bird to the show in a good-sized travelling box with a thick layer of shavings in the base. You do not want to waste all your hard work and risk broken feathers by using a cramped box. Avoid travelling boxes with bars, as these can ruin the tail in next to no time. A solid cardboard box with air vents is often the best alternative to a purpose-made box.

Arrive in plenty of time to do any last-minute tidying of the birds. You will have been given a penning slip, with the numbers of the pens in which to place your birds. Before penning the bird, add a small smear of petroleum jelly to the legs and face of the bird to make them shine, then settle the bird in the show pen to await results.

To prevent the birds soiling their feathers, it is customary not to feed or water the birds until after the judges have finished their rounds. A full crop can distort the overall shape of the bird, especially in gamefowl.

Do not talk to the judge while he is examining the birds, but sit back and await the decision with the rest of the exhibitors. Hopefully, you will find a red card on your pen, telling you that you have won first prize. If not, then do not be shy to ask the judge, after he has finished judging, to point out why he marked down your bird. It is all part of learning the show scene. You will also find other experienced exhibitors will readily offer advice.

After the show

You are not permitted to 'box' your birds to remove them from the show hall until after a specified time, which will be stated in the schedule. There may be a short wait before you can actually leave the hall to ensure that all the birds are accounted for. If you have time at the end of the show, do offer to help take down a few pens, as the show organisers, who

are all volunteers, are frequently left to cope with this task with very little assistance.

On returning from the show, treat your birds for mites, just in case they picked up any at the show from a neighbouring bird whose owner was not as scrupulous as you were. For biosecurity reasons it is always best to keep the birds quarantined from the main flock for about two weeks after returning from the show.

Above: The red card on this pen denotes that this Orpington has won its class. Winning at a show makes all that effort and care seem very worthwhile.

81

Introduction to Light Breeds

The birds that fall under this category are the laying breeds, many of them descended from the farmyard stock that provided eggs for the table throughout Europe and beyond for thousands of years. Refinement of the breeds resulted in fowl that were good at converting feed into large numbers of eggs, rather than flesh, meaning that these birds did not waste energy putting meat on their bones. Thus, they are smaller-framed, with finer bones than those fowl required for the table. All the commercial hybrid egg producers found today were originally based on these light-bodied laying breeds.

Right: The Welsummer is one of the largest of the Light Breeds, and known for its dark brown eggs.

The process of going broody and raising a family would halt egg production for many weeks, so selection was made away from this trait. The result is that today, very few of the Light Breeds are known to go broody – and if they do they soon stop and

resume laying, rather than completing the act. If you wish to breed from Light Breed fowl you will need to use an incubator. Some of the more fancy-looking breeds, such as the Sultan, also fall into this category.

The Light Breeds are far more alert and active than other types of chicken and in general are rather more flighty and nervous than the Heavy Breeds. Many of the Light Breeds readily take to roosting in trees; indeed, some have excellent flying abilities. Once they were developed in bantam form, the light body and long wings resulted in birds that can easily fly considerable distances if they choose to.

Right: Some of the smaller-bodied Light Breeds, such as this Fayoumi, can almost be mistaken for a Bantam when viewed against some large Heavy Breeds.

Light Breeds are usually not as long lived as Heavy Breeds, but if you are looking for a pretty fowl, and the number of eggs is important to you, then you should look amongst the Light Breeds for your choice.

Ancona

This breed was originally common around the Ancona area of Italy. Early books apparently reported any black-and-white birds from this area being given the name Ancona. However, from descriptions of the Ancona fowl exhibited at the Birmingham, UK, shows held in the late 1850s, it seems that the birds closely match those that we now know as the true breed. A Bantam version of the Ancona was developed around 1900 and is now seen more frequently than the Large Fowl.

There is no doubt that the Ancona bears an exceedingly close historical relationship with the Leghorn, with many people claiming it was the original mottled Leghorn.

Appearance and temperament

The Ancona is a very attractive breed, active and alert, but not large. On average, the males weigh around 2.7kg and the hens around 2.25 kg.

The plumage is black, with a beetle-green sheen to it. Each feather should be spangled, with a V-shaped marking of white on the tip. This white spangling will become more pronounced after each moult. As well as the single-combed variety there is a rose-combed type, but it is less common. Like other Mediterranean breeds, the birds sport white earlobes.

In common with most Mediterranean breeds, the Ancona is not overly friendly, but will normally respond to regular handling. It is suitable for a novice keeper.

The Ancona is well regarded as a good laying fowl. The eggs are large and chalk-white in appearance; it is one of the better producers amongst the pure breeds.

Points to bear in mind

Top show breeders will sometimes use double mating to achieve the required comb folds in the single-combed variety, but this is not always necessary.

Below: *There should be an even V tip of white to the feathers in Ancona. This becomes more pronounced after each moult.*

Andalusian

As it closely resembles the Minorca and the Spanish in shape, early poultry fanciers frequently believed the Andalusian to be a subvariety of these breeds. Early types imported into the UK in about 1846 from Andalusia showed a large variation of feather pattern, but the breed was carefully selected over the following years for the well-defined lacing seen today.

Appearance and temperament

The colour is always the most striking feature of the Andalusian. The breed met with some early opposition, as confusion arose over the way the colour appeared unstable. It was discovered that crossing two blue birds results not in all-blue offspring, as might be expected, but in 50% blue chicks, 25% black chicks and 25% chicks with an off-white colour called Splash. If the Splash colour is subsequently crossed with the Black, then all the chicks are Blue (but of varying degrees of perfection).

We now know that the number of blue genes the bird carries determines three possible colours: two copies of the blue gene, result in a Splash colour; one copy results in the Blue colour; and no copies result in the Black colour. This breed exhibited one of the earliest examples of a gene showing incomplete dominance ever recorded, and is now widely used to illustrate the inheritance laws of Mendelism.

Breeding these birds to show standard is an art and novices are advised to start with a stable colour first. As only the Blue form is permitted to be shown, it means that many unwanted Black and Splash Andalusians are produced.

The earlobes are white, the neck hackle a lustrous black and the feather pattern should be an even shade of blue-grey, surrounded by crisp black lacing. This is very difficult to achieve.

The medium-sized Andalusian should be a robust, active breed that lays large chalk-white eggs. It does not tend to go broody.

Points to bear in mind

The breed should not pose any major management problems. Take care over the white earlobes, which tend to blister if damaged.

Although slightly flighty, Black and Splash hens make reasonable layers for the back garden and the eggs are usually a good size.

Below: The Andalusian is a suitable breed if you want laying birds for the garden, but breeding them to show standard is very difficult.

Appenzeller

Finding information about the early history of this breed is difficult, but it appears that the Gold variety is very similar in appearance to the now extinct Yorkshire Hornet. In an article in the 1981 edition of the *Poultry Club Yearbook*, Pamela Jackson describes how she imported this already rare, old, Swiss breed into the UK and successfully hatched chicks in 1973. She mentions that the breed may already have been in the UK in the 1940s, as she had found mention of one at a poultry show of that time.

Appearance and temperament

The Appenzeller Spitzhauben is named after the lace bonnets worn in the Appenzell region of Switzerland. The breed is found in three colours: Silver Spangled, Gold Spangled and Black. The Silver Spangled is the most frequently encountered. This striking breed has a forward-facing crest, a horn-shaped comb and large, cavernous nostrils.

The Appenzeller is a small, exceedingly active, alert and hardy breed, and happiest when provided with plenty of space, rather than being enclosed in a small pen. A Bantam version is now available.

Males can prove very virile and determined when seeking out available females. If you keep more than one breed on the premises, be vigilant, otherwise every chick in the area will hatch wearing a cute little crest!

The Appenzeller is an early-maturing breed; the hens come into lay at about five months and provide a good supply of small white eggs.

The Appenzeller Barthuhner is a heavier form of the breed, with a beard and a rose comb but no crest. It is not as popular as the Spitzhauben.

Points to bear in mind

Appenzellers are a very flighty breed and need plenty of handling from an early age if they are to become tame. They are very prone to roosting in trees if the opportunity arises.

Left: The nostrils are large and cavernous and the eyes alert.

Below: The horn comb has no side sprigs, the horns being equal length.

The crest is said to resemble a lace bonnet.

Left: Black tips to the tail feathers are a distinctive feature of the Appenzeller. Novice keepers are recommended to try a more docile breed first.

Araucana

The blue-egg-laying, rumpless South American fowl called the Colloncas and the ear-tufted Quetros are almost certainly the ancestors of the breed known today as the Araucana. Originating in Chile, the Araucana takes its name from the Arauca Indians. In 1914 Professor Salvador Castello brought Araucanas to the attention of the Western World. He speculated that the blue egg gene may come from a wild fowl called the Chachalaca, which reportedly has occasionally hybridised with domestic poultry.

Right: The rumpless version without a tail. Note the unusual ear tufts on the rumpless variety.

Appearance and temperament

The Araucana appears in a number of forms. In the USA, the rumpless version with unique ear-tufts that spring from a fleshy pad beside the ear is accepted as the true breed.

Unfortunately, the gene for these ear-tufts carries factors resulting in many chicks dying before they hatch. Of those that do hatch, a number will have lopsided ear-tufts. Rumplessness also gives fertility and management problems; all this makes the rumpless form exceedingly difficult to breed. Therefore, poultry enthusiasts developed new forms of the breed that still laid the desired blue eggs but reproduced more easily. The British developed a tailed version of Araucana and it became standardised in 1969. In the USA, a tailed version, known as the Ameraucana, was accepted as a standard breed in 1984.

The Ameraucana and British Araucana look very alike, the main difference being that the British version has a crest. Both these more recently created breeds have muffling around the face instead of ear-tufts and no known problems with fertility or hatching.

The Araucana is always pea-combed. It is found in a range of colours, with the Self-blue colouring called Lavender being the most popular.

Points to bear in mind

The tailed version of the Araucana and Ameraucana should cause no problems for novice keepers; just keep a check for mites in the crest and muffling. Novices should avoid the rumpless ear-tufted version until they have acquired more poultry-keeping experience.

A male Black-Red British Araucana, showing a crest and muffling around the face.

Ardenner

The Ardenner is widely regarded as the oldest of the Belgian breeds. Little is known about its distant origins, but there is no doubt that it has been a very practical farmyard fowl in the Ardennes region for an exceedingly long time.

In 1893 a poultry association for the promotion of the Large Fowl Ardenner was set up in Liège, and in 1913 a Bantam version was standardised.

Appearance and temperament

The breed should be mulberry-faced, meaning that it shows a darker skin pigment than the bright red usually found in poultry faces.

The Ardenner is one of the few breeds also found in a rumpless variety, with a total lack of tail vertebrae (parson's nose). The poultry writer Edward Brown mentions that these rumpless types, known as Hedge Fowls, were popular in the Liège district and kept on isolated farms.

Above: Ardenner male. Judging by its appearance and wild nature, it was thought to have descended from "a race bred under rugged conditions, crossed with one or other type of Game fowl".

Being a very active breed, the Ardenner is miserable unless permitted free range. It really enjoys foraging over a very wide area and will readily fly over the usual height mesh of a poultry run and roost in trees.

As is the case with many of the European breeds, the World Wars caused a great loss of numbers. The breed is still very scarce and locating suitable available stock will be difficult.

Points to bear in mind

The Ardenner is not a naturally friendly breed and dislikes being handled. Children wanting a hen as a pet will be very disappointed by the temperament of this breed. Not suitable for a novice. They are a free range fowl only, and even then you will spend a great deal of time retrieving the birds from your neighbours' gardens. Choose a tamer breed.

Above: A rumpless Ardenner hen. The face colour of the Ardenner should be darkly pigmented, but this is far more apparent in the females than in the males.

Autosexing breeds

Autosexing breeds are so named because it is possible to sex the chicks by down colour on hatching.

It was Mr. M. Pease and Professor Punnett who discovered that the barring gene is closely linked to the sex of the breed. A double dose is found in the male and a single dose in the female. When combined with a brown gene, the change in the down colour of the chicks becomes very noticeable, with the male chicks being much paler in colour than their sisters and thus easy to sex on hatching.

Breeds carrying these autosexing genes were subsequently developed. All of them can be identified by the letters 'bar' at the end of their name. As the name suggests, all these breeds should carry some form of barring in their plumage.

Above: These two Welbar pullets would have been distinguishable from their brothers by having a far darker down colour when they hatched.

Unlike the crosses between gold and silver, which usually involve two different breeds, the autosexing breeds are true pure breeding fowl that need no outcross to show the trait.

The first breed produced was the Cambar in 1929, derived from the Campine and Barred Plymouth Rock. It was then recrossed back to utility Barred Rocks. Very few of these are still in existence.

The Legbar was developed using Leghorn and Barred Plymouth Rocks and is found in three colours: Gold, Silver and Cream. Adding Araucana genes to the cream version produced an autosexing breed that lays blue eggs. This latter breed is still found in reasonable numbers.

Above: The tail of the Cream Legbar should be evenly barred, although some extra white is permissible, and the sickle feathers are often paler in colour.

The Brussbar was created using the Brown Sussex and designed as a good, all-round utility fowl, providing both eggs and meat.

The Rhodebar was produced using the Brussbar and the Rhode Island Red, again as an autosexing utility breed. They had died out, but have recently been recreated and are starting to grow in popularity. The red barring is very attractive.

The Welbar, found in the Gold version only, was developed from the Welsummer and Barred Plymouth Rock to create a useful utility breed that lays brown eggs.

Points to bear in mind

Obtaining genuine stock can be a problem. Many people try crossing these breeds haphazardly, thinking they will all be autosexing, but this is not so. Parent strains need careful selection. Beware of commercial strains of the Cream Legbar, as these do not lay blue eggs consistently, or display reliable autosexing properties. Seek out a reputable breeder. All autosexing breeds are very suitable for novices and should give no problems.

Cream Legbars have a small cream and grey crest.

Above: *An infusion of Araucana blood during the creation of the Cream Legbar resulted in the breed producing pale blue or green eggs.*

Bassette

The Bassette as a standard breed was created from birds of good laying abilities found around the South-Limburg and Liège region in Belgium at the beginning of the twentieth century. Although the name Bassette had been used to describe the local poultry population many years previously, it was Mr. W. Collier who standardised the breed as an excellent small laying fowl, first exhibiting it in 1917.

The breed very nearly became extinct after the Second World War. The few remaining birds were brought together at Gendt University and the breed was saved. However, locating available stock may still prove problematical.

The Bassette is not yet standardised in the UK, so is included in the Light Breed section of this book, as that is possibly where it will be classified. However, as no truly Large Fowl is in existence, it could just as well have been placed in the True Bantam section.

With the more open borders in Europe, it is inevitable that this breed will very soon make its way to the UK,

if it has not already done so, especially with the increasing demand for small utility-type birds. It is a hardy, tough and productive little bird, ideally suited to a smallholding or similar environment.

Appearance and temperament

The Bassette stands as a half-size breed, part way between a Large Fowl and a Bantam, and is often called a Semi-Bantam. It is found in a number of colour varieties; Quail and Silver Quail are the most often seen. A Lavender Silver-Quail colour is unique to this breed.

Bassette are very active fowl, doing best where they can forage freely. Closely confined birds can be prone to develop vices unless kept occupied. Be aware that the breed can fly well and will take to roosting in trees if the opportunity arises. They should cause no real problems for the novice.

Right: The Bassette is a small-sized, practical, utility breed, with an active, alert temperament.

Brabanter

Breeds with beards and crests have been mentioned since the very early years of selective poultry breeding. Some of the earliest evidence of the Brabanter can be found in the paintings of the Dutch artist Melchior d'Hondecoeter. His painting *A Hen with Peacocks* clearly shows a hen very similar to a Brabanter, with a forward-facing crest and very pronounced beard. She must have been a favourite subject as he painted the same hen again in *Birds and a Spaniel in a Garden*. This time he added the cock bird, which leaves little doubt that these were early Brabanters. Melchior d'Hondecoeter rarely dated his pictures, but they are attributed to the mid-1670s.

Above: *Note the unusual 'trilobed' beard. Regularly check both the beard and the crest for signs of mite infestation. As the beards can become soiled, always provide water in founts rather than open bowls.*

Once breeds such as the Wyandotte, the Orpington and, in particular, the Leghorn arrived on the scene, the Brabanter rapidly fell out of favour. It was soon edging towards extinction and only saved by a few devoted fanciers. Using the last remaining birds in existence, they resorted to outcrossing to its close relative, the Dutch Owlbeard, in a bid to retain at least some of the gene pool. So it was that by the 1920s a 'new' form of the breed arose, which was somewhat smaller than the original.

Appearance and temperament

The Brabanter is a medium-sized breed and one of its most notable features is the forward-facing crest, similar in appearance to that of the Appenzeller. It has a beard that is described as trilobed, and muffling around the face. As with most crested breeds, it has a duplex horn-shaped comb. Popular colours include the Gold Spangled, Silver Spangled and the Chamois.

The hens are reported to be very good layers.

Points to bear in mind

The beard and crest can be very prone to mites, and must be checked on a regular basis. In the UK, finding stock will be difficult, as the breed is very rare and numbers vary depending on intermittent imports. Brabanters are suitable for the novice with some extra care.

Brakel

This fowl was bred extensively in the region of Nederbraekel in Flanders. Reports indicate that it was already known there as early as 1400.

The Brakel is a rare breed, snatched from the brink of extinction last century. From the foundation of the Brakel club in 1898, this once well-regarded breed sank into obscurity after the Second World War. By 1971, all that could be found of the breed were two hens, two cocks and about a dozen eggs. Amazingly, the breed was revived from these tiny numbers and in Belgium it is now fairly popular. Additional colour varieties have been developed, including the very attractive Citroen and Chamois.

Appearance and temperament

The Brakel is very similar to the Campine in appearance, but the Brakel is heavier, weighing around 3.2kg. It is stockier, with a deeper-shaped body than the Campine, and carries a full and well-furnished tail and neck hackle.

The colouring is very attractive, with each feather barred in a transverse direction. This results in the bird looking like someone has drawn rings around it.

The eyes of the Brakel should always be as dark as possible. It should have evenly shaped, medium-sized, oval, white earlobes and slate-blue legs. The

Above: Look for a good, full tail in the male and good saddle furnishings. Strains that have been outcrossed with the Campine tend to be shorter in the tail feathers. This picture clearly shows that the Brakel is a heavier type than the Campine.

neck hackle of the male should be a clear colour, with no obvious ticking or spotting.

As a breed, it tends not to go broody and is a good layer of reasonably sized, white eggs.

If available stock can be located, this rather flighty breed should pose no problems for the novice.

Breda

The Breda has never been a hugely popular breed, possibly due to the fact that it came to attention at around the same time as popular breeds such as the Cochin, Brahma and Leghorn. For a while prior to the American Civil War (1861) it was kept in reasonable numbers in the USA, but faded from sight soon after.

In *Harpers New Monthly Magazine* of 1873, the utility qualities of the breed merit a passing mention: "The hen has but little attraction to the ennui of incubation – the fault of all good layers that fatten easily." The breed is also mentioned in Moubray's *Treatise on Domestic and Ornamental Poultry,* where he states that the birds are called Guelderlanders.

In the Netherlands the breed is known as Kraaikoppen which translates as 'crow headed'. Much confusion arises between the Dutch name of this breed and the name Kraienköppe used in Germany. The latter refers to a totally different breed called the Twentse in the Netherlands (see page 100).

The Breda is now only kept in small numbers, even in its homeland of the Netherlands.

Appearance and temperament
The Breda is unique amongst poultry as it is the only breed that naturally has no comb. This is caused by a variation of the duplex comb gene, such as that found in the La Flèche. Consequently, it has large nostrils, something found in all breeds with duplex combs. This comblessness has always made the Breda a popular fowl with geneticists wishing to study the heredity of comb types.

A Bantam form was developed around 1930 and is available in Black, Blue-Laced, White and Cuckoo. The breed is said to lay well, mature fast and quickly become tame, making good pets.

Points to bear in mind
Maintain a lookout for scaly leg and keep the breed out of the mud. Provide extra perch space to cater for the foot feathering. Suitable for a novice, but locating stock will be difficult.

Below: The Breda naturally has no comb, although a rudimentary comb can often be seen in the form of just a couple of small pimples.

Campine

The Campine originated in the Campine region of Belgium, which includes the eastern areas of Antwerp, as well as parts of the Netherlands province of North Brabant. The breed is mentioned in the USA Commissioner of Agriculture's report of 1866, although there he refers to them as a Dutch breed. From its appearance it is very obvious that the Campine shares much of its origin with the Brakel (see page 92).

Appearance and temperament

About a century ago, British breeders decided to distinguish the two breeds fully by selectively breeding the Campine for a trait known as 'hen feathering'. This means that the cockerels fail to develop the usual male distinguishing features of large curved sickle feathers and abundant saddle and hackle feathers, and retain a more henlike appearance.

Below: The Campine is suitable for a novice, but try a friendlier breed first. These chickens are good layers and very active foragers.

If you look at a male Brakel, you will see plenty of white or gold saddle feathers covering the back of the bird. However, in the Campine, the barred colour stretches all the way back to include the much shorter tail. The barring should not be of an even width; the black bars should be three times the size of the white or gold that separates them. The Campine is normally found in the Silver or Gold colours, with a solid coloured background covered with black barring.

When choosing a Campine, look for a distinct lack of saddle feathers in the male, dark brown eyes, medium-sized white earlobes and blue legs.

Hen feathering can often cause fertility problems, although the Campine appears to have retained its vigour over the years. Campines are good layers of white eggs, excellent foragers, fast maturing and early feathering. Like all Light European breeds they are active birds, best permitted plenty of space. In common with many old pure breeds, they fell out of favour with farmers once the commercial hybrid fowl arrived on the scene.

Points to bear in mind

The Campine should give no particular problems, but it is not a friendly bird. It would far rather be out and about behaving like a natural chicken than be regarded as a pet.

Dandarawi

For those who like the challenge of something with a touch of the feral about it, this may be the breed for you. It is classed as a rare breed, with only limited numbers available at present, and competes in the non-standard classes.

The Dandarawi is a recent addition to the UK and European poultry scene. Hailing from the Egyptian town of Dendera on the West bank of the Nile, the breed remained unchanged for centuries and stayed isolated to this one area. The Dandarawi is one of the native laying strains of Egypt. It is a hardy breed, not prone to disease, and a good layer of numerous small white eggs. Potential breeders should remember to concentrate on keeping the breed for its character and laying ability.

Below: The unusual cup comb is a feature of the Dandarawi and a trait that it shares with the Sicilian Buttercup.

Appearance

The Dandarawi has an unusual cup-shaped comb, which hints at a common ancestor with the Sicilian Buttercup (see page 111). In addition, it can also sport a very small tassel of a crest.

Although classed as a Large Fowl, the Dandarawi is a small breed of chicken, with no Bantams yet available. The chicks are reportedly easily sexed at hatching as the females show black head spots. The breed matures very fast and the cockerels have been known to crow as early as 25 days old.

Temperament and behaviour

The Dandarawi is an intelligent and alert breed, with a strong self-survival instinct. It is exceedingly active and can fly very well. You will need a run with a roof to prevent the birds taking up part-time residence in the surrounding trees. In attitude and temperament the Dandarawi is naturally rather wary and suspicious, but with time and persistence it will become reasonably tame. However, children will be disappointed if they expected to have a friendly pet. A novice keeper would be well advised to try a more docile breed first.

Below: A newcomer to the UK, the Dandarawi is already gaining admirers.

A Wheaten coloured Dandarawi hen.

Fayoumi

This very old breed was first brought to the attention of the Western World by Mohammed Askar Bey, when he presented a paper called *Poultry Keeping in Egypt* at the 1927 World Poultry Congress. He mentioned that the breed from the City of Fayum, having a fixed colour and type, would be worthy of being recognised as a distinct breed. He continued by saying that locally the breed was believed to have been of Turkish origin. The fact that it resembles the Campine in plumage pattern may indicate a common ancestor.

The Fayoumi has remained as true to its original wild ancestors as it is possible to find. In its native country it has only ever been bred for ruggedness and egg production. It has adapted well to changes in climate and is now found in many countries.

Above: *A male Fayoumi. This breed is always alert and ready for action if the need arises.*

Appearance and temperament

This is a small breed, with a fully adult male only weighing around 1.8 kg. No bantam form is available yet.

It takes an effort to become real friends with one of these exceedingly alert, very active and wary birds. However, once you do, it is intelligent enough to tame down sufficiently to be handled and shown successfully, taking to confinement without too much objection. If handling ceases, it will quickly revert to its wary and wild nature. It is not suitable as a pet breed.

The Fayoumi matures early and is a very good layer of small white eggs. It flies exceptionally well, which enables it to be fairly adept at avoiding many predators. Roosting in trees and active foraging is second nature to these birds, so you will need a pen with a netted roof if you wish to prevent them straying. Not suitable for a novice unless you enjoy a challenge. Try a tamer breed first.

Right: *The Fayoumi matures early and pullets will sometimes lay as early as 18 weeks.*

Friesian

The Friesian is one of the oldest breeds from the Netherlands and one of the basic productive farmyard breeds of Europe. It is low on feed consumption, an active forager, alert to predators and a very good egg layer of small white eggs. It matures early and seldom goes broody, resulting in fairly constant egg production.

Appearance and temperament

The Friesian is a small bird, the diminutive male Bantam only weighing up to 650gm (females 550gm). The Large Fowl version is often mistaken for a Bantam, as it weighs only about 1.6kg.

In its homeland and also in Germany, the breed is found in a number of attractive colours. Elsewhere, the most frequently found colour is the very pretty Chamois Pencilled, with delicate white pencilling on a bright pale-gold ground colour seen in the females. Males are a bright golden buff colour, with white tails.

A Friesian is always active and alert and happiest when given extensive free range. Be aware that if you do need to keep it in a confined run, it will need a wire or netted roof to prevent it flying out. Like most of the Light European breeds, it is equipped with strong

wings and knows very well how to use them. Given the opportunity, this breed is very prone to sleeping out in the trees – something your neighbours might not appreciate when a cockerel takes to crowing outside their bedroom window in the early hours of the morning!

Points to bear in mind

This is not a breed for people who want a cuddly pet. Children will be disappointed if they expected tame, friendly hens. Apart from that, it is a vigorous, hardy, very ancient breed.

Below: The Friesian is a good laying breed, producing a large number of small white eggs. This very flighty breed is not really suitable for novices. Try a more docile breed of poultry first.

Hamburgh

The Hamburgh is of North European origin and it appears that the poultry writer Aldrovandi was familiar with the breed as far back as 1600. A bird he refers to as a Turkish hen sounds exceedingly like a Spangled Hamburgh, when he says she is "all white, sprinkled over with black spots; she has the feet tinged with blue and the wattles short when compared with those of the male." Another hen presents the same appearance, except that she carries a "sharp point on the top of her head" – almost certainly a reference to the rose comb that was later adopted as the standard.

It was in the early years of the nineteenth century that competitive farmers in Lancashire and Yorkshire started to stage exhibitions and compete for prizes of household goods and even developed a detailed written standard for the breed.

Early pictures from around 1846 show birds with crests, but by 1860 John Baily, writing in *Fowls; A Treatise on the Principle Breeds*, gives a description that would be recognised by fanciers today, stating that no true Hamburgh should have a topknot or a single comb.

Above: *To breed hens showing good pencilling, use a male bird that carries a reasonable amount of visible pencilling on his body.*

The spelling of this fowl's name is only one of the minor controversies this breed has caused over the years. In the UK, the name is spelt with an H on the end. It appears that this name was coined for the breed by the Rev. E. S. Dixon in 1848.

Writing in 1890, H. S. Babcock says of the Hamburgh "It has masqueraded under a great variety of names such as Pencilled Dutch, Everlasting Layers, Dutch Everyday Layers, Bolton Grays, Creoles, Corals,

Left: *In comparison to the Pencilled hen above, the show-quality Gold Pencilled male in this group shows a clear body colour.*

Right: *The Silver Spangled colour always attracts the most admiration. Note the perfect contrast of black and white in the tail feathers.*

Creels, Bolton Bays, Silver Pheasants, Silver Mooneys, Silver Moss, Golden Pheasants, Golden Mooneys, Copper Moss, and the like; but, under whatever name, it has always won a host of admirers."

Appearance and temperament

The Hamburgh is standardised in five colours of which the Black is almost certainly now the rarest. Although not officially classified as a rare breed, because it has its own breed club, the Black Hamburgh is so scarce that unless more breeders are found it may soon disappear completely.

The other four colours standardised are Silver and Gold Spangled and Silver and Gold Pencilled. The Silver Spangled is the most frequently encountered of these colours. In the USA and on the European Continent, colours such as White and Blue are also available. Bantam versions were developed in around 1900.

The Pencilled variety was perfected in the Netherlands. Originally, these birds were smaller in size than the other colours, but are now similar. This colour needs to be double mated (see page 76) to produce show standard birds, as the plumage pattern of the male and female are very different and cannot be produced from a single breeding pen.

Potential breeders must study the standards of the Hamburgh very carefully, as there is far more to getting a Hamburgh right than just having a round spot on the end of the feathers or good-quality pencilling.

Being a Light Breed, Hamburghs are often rather flighty, although this can depend on the individual strain. However, they are slightly less scatty than many of their cousins and make very good show birds. Nevertheless, novices may like to try a tamer breed first.

Hamburghs are reliable layers, producing a good number of white eggs.

Points to bear in mind

A covered run is advisable as Hamburghs may take to sleeping up trees, but otherwise they should present their owners with few problems.

Kraienköppe

This breed (also know as the Twentse in the Netherlands) should not be confused with the Breda (see page 93) that in the Netherlands goes by the name of Kraaikoppen.

Although developed on the German/Dutch borderland during the latter part of the nineteenth century, the Twentse breed was not shown in the Netherlands until 1920. It was around this time that the addition of Silver Duckwing Leghorn resulted not only in better egg laying ability, but also the colour that is most usually found today.

Appearance and temperament

The Kraienköppe (pronounced Kry-en-cope-ah) is a very distinctive-looking breed. Indications of its Malay Game ancestry can be clearly seen in the male's strong head features, namely overhanging brows, small walnut comb and almost absent wattles. The birds have an upright stance and very full, flowing and open tails.

The Kraienköppe is not generally a very friendly breed. It can be rather nervous and, on occasion, decidedly temperamental, especially when being shown. It does not settle well in confined areas, relishing the ability to forage and free range.

In spite of looking like a game fowl breed and having Malay Game in its distant ancestry, the Kraienköppe is reportedly lacking in fighting instinct and is no more aggressive than any other Light Breed male.

This breed reputedly lays well during the winter months, providing you can stop the hens going broody – they regard this as a very regular way to pass the time. If allowed to hatch chicks, the hens usually make very good, if rather fierce, mothers.

Points to bear in mind

Unfortunately, although the Kraienköppe is not an aggressive breed, its very gamelike appearance tends to draw it to the attention of people who steal game fowl for illegal cockfighting. For this reason alone, the breed really cannot be recommended for novices. However, the Bantam version would attract less interest and could be considered.

Left: The narrow walnut comb should look rather like an elongated strawberry. The males have very long, flowing tail feathers.

Leghorn

The Leghorn takes its name from the port of Leghorn in Italy. The first real documented evidence we have of Leghorns is in 1835, when a consignment of Brown Leghorns was imported by N. P. Ward of New York, USA. It appears that he very quickly distributed hatching eggs amongst his friends.

In the USA, breeders set about encouraging the breed's undoubted laying ability, and by the time the first White Leghorns arrived in the UK in 1870, they were already well regarded as a superb laying breed. The Brown Leghorn followed two years later.

The town of Petaluma in the USA based such a vast industry around the commercial properties of the Leghorn, that by 1916 it held its first Annual Egg Day to celebrate. It would be safe to say that all modern commercial laying hens are still extensively based around the Leghorn breed.

Appearance and temperament

It is important to note that the breed of Leghorn in the UK and the breed found in the USA parted

Above: *The American Leghorn is a different shape to that of the UK version, having a smaller comb and larger tail.*

company in appearance a long time ago. The American Leghorn still retains a similar appearance to the original productive utility strains found prior to the First World War, with a large tail and medium-sized comb. The type now found in the UK has a shorter tail and a very large comb, with much exhibition emphasis being placed on head points.

The Leghorn is an active bird, seldom going broody, and if you acquire utility stock will provide a good supply of white eggs. It is a very useful back garden breed and does well in a more confined environment. The Brown Leghorn, in particular, retains a smart appearance throughout much of the year.

Points to bear in mind

For showing, top breeders may use double mating, but this is not always necessary. As with all Light Breeds Leghorns can be rather flighty, but they are an intelligent bird and can even tame down enough to make reasonable pets.

Below: *The plumage of the Brown Leghorn tends to stay looking clean and smart throughout the year.*

Lakenvelder

The Netherlands and Germany both lay claim to being the originators of this striking-looking breed. The Dutch claim that the breed was mentioned in a travel journal of around the 1700s and that it takes its named from the village of Lakervelt in the Utrecht region. However, the Germans claim that the breed was developed in the Westphalia area of Germany during the eighteenth century. The breed was certainly well documented in Germany by 1835.

What is clear is that the two countries have slightly different standards for the breed. The German version has a clear white back, while the Dutch version has a blacker saddle and the undercolour is also a darker shade. The UK standards for the breed are based on the standards produced in the Netherlands. Hence, the breed name is spelt the Dutch way, as Lakenvelder, rather than the German spelling of Lakenfelder.

Appearance

The bird should have white body feathering, with pale grey underfluff and blue legs. Look for a clean black neck hackle, free from

Right: The Lakenvelder colour is sometimes described as 'a shadow on a sheet'.

white feathering. A Blue variant in the Bantams, in which the black part of the feathering is replaced with an even pigeon-blue, was developed in the Netherlands a few years ago and now birds of this colour occasionally turn up at shows in the UK.

Temperament and behaviour

The Lakenvelder can be rather highly strung and often the first breed to sound the alert to any real or perceived danger. It is not naturally a tame breed and although regular handling will improve matters considerably, birds can frequently be a bit nervous when shown. The breed tends to be a naturally light sleeper and will crow if woken by any disturbance during the night. Novices are advised to try a less flighty breed first.

Large Fowl males are generally very good tempered throughout the year, but the Bantam males can be a bit naughty during the breeding season and are best kept separate from children at this time.

The Lakenvelder Bantam is superb at flying. It is able to take off like a vertical take-off jet plane when the need arises. This may be to escape a predator or possibly just to fly to the far

Below: *It is unusual to find a neck hackle that does not show some white just behind the head.*

of medium-sized white eggs, with the Bantams frequently laying more eggs than the Large Fowl.

Breeding

It cannot be said that the Lakenvelder is an easy bird to breed. To obtain the correct colour pattern involves a great deal of careful selection of the parent stock and even then, producing perfectly marked offspring from these parents is not a guaranteed certainty.

It is impossible to tell at an early age which chicks will mature into the best-coloured birds. This means keeping the chicks until they have nearly acquired their full adult feathering before making a selection. Often, the birds with the best pale grey undercolour will produce black speckling in the top colour, or the birds with the best top colour in the body will have undesirable white feathers showing up in the neck hackle.

Producing top-quality show birds is an uphill struggle, but their beauty can more than make up for the problems encountered.

end of the garden to avoid walking. Sleeping in trees is a very popular pastime with the Bantams, so covered runs are vital if you wish to prevent them straying.

Although very good natural foragers, the Large Fowl are not a particularly destructive breed. Six Lakenvelders will do a lot less damage to a planted enclosure than half that number of many other breeds.

The Lakenvelder is a non-sitting breed and in many years of keeping it the author has never yet known a Lakenvelder to make any attempt at going broody. The birds are reasonable layers

Above: *A Blue Lakenvelder male. Crossing the Blue and standard coloured Lakenvelder will produce a percentage of chicks of each colour.*

103

Marsh Daisy

The history of this breed starts in 1880, when John Wright of Marshside, Lancashire, UK started to cross various breeds. A mix between White Leghorn, Malay, Black Hamburgh and Old English Game proved to be a very useful sort of fowl for his purposes and he kept the birds as a closed flock, without adding in any other breeds, for 30 years. During this time it became very uniform in type.

In 1913, Charles Moore from Doncaster bought two of the remaining hens from John Wright. He mated these hens to a pit-game cock and then later added Sicilian Buttercup. A number of other breeders took up these fowl and by 1922 they were admitted to the poultry standards under the name of Marsh Daisy. They fared well for a number of years but then faded from the scene and were thought to have become extinct. However, a flock was discovered in Somerset in 1971 and the breed was revived.

Appearance and temperament

The Wheaten colour is most often seen, although Buff and Brown are also available. Look for white earlobes and willow (greenish) coloured legs.

The Marsh Daisy makes a good smallholder's breed, being tough, hardy and long lived. They should pose no problems for novices. The hens lay a reasonable number of smallish eggs and the males carry a fair amount of flesh for the size of the breed. They are very good at foraging and are best kept where they can have extensive ground to roam over. Contrary to popular belief, the Marsh Daisy is no better at living on marshy ground than any other breed of clean-legged, robust poultry. If you wish to keep fowl on marshy ground, you would be better off with ducks.

The Marsh Daisy is still very rare, making only occasional appearances at shows and then only in low numbers. The breed does have its devotees and stock is sometimes advertised in the poultry press.

Left: *The Marsh Daisy makes a good choice for a novice wishing to embark on keeping rare breeds.*

Minorca

The British spent most of the eighteenth century fighting over dominion of the Island of Minorca (Menorca) and it is supposed that it was some time during the latter part of that century that the Minorca was first taken back to the UK. However, it is not until 1834 that we find documented evidence of the breed. It is now almost extinct on the Isle of Minorca, but some enthusiasts on the island are trying to revive and preserve the breed as it originally was.

The Minorca was refined by British breeders into an excellent laying breed that by the late 1800s was known the world over for its large white eggs. Although it never reached the same level of popularity as breeds such as Leghorns, the Minorca did well in backyard and confined environments and was a very popular smallholder's breed.

The Bantam was developed by the late 1800s and also proved to be an excellent layer.

The rise of the commercial hybrid eclipsed the Minorca for laying ability and the breed fell from popularity, but is still kept by enough fanciers to ensure its future.

Appearance and temperament

The breed is found in Black, Blue, White and Buff. A champagne-blonde colour mutation was once in existence, but now appears to be extinct. The large oval-shaped earlobes are a distinctive feature.

The Minorca is not as flighty as many Light Breeds, with the Bantam hens, in particular, being very opinionated and bossy little madams.

They are well noted as a non-sitting breed, and although this is still true of the Large Fowl, it appears that the Bantam version, at least, is unable to read and therefore cheerfully ignores this documented fact!

Points to bear in mind

The earlobes are prone to blistering, so take good care to protect them from damage, as it takes a while for them to repair. Show breeders will use double mating to ensure the correct comb type in males and females (see page 76).

Left: *This pullet's comb will grow larger and fall over to one side as she matures.*

Left: *The earlobes of the Minorca should be oval in shape. They blister very easily if damaged.*

Old English Pheasant Fowl

This breed officially came into being in 1914 as a result of the work of certain fanciers who were endeavouring to maintain the older type of the Yorkshire and Lancashire Pheasant. (Over many years, that breed had been moulded into the popular Hamburgh, with its perfect feather pattern and high exhibition standards.)

The Old English Pheasant Fowl (often shortened to OEPF when written down) has remained a dual-purpose breed since its standardisation. Although classed as a light breed, adult males should weigh around 3.2kg, so spare males are worth rearing on as table birds even if they will not match up to today's modern meat breeds in the quantity of flesh they provide.

Appearance and temperament

The rose comb should be of moderate size, with the leader following – but not touching – the line of the neck. The ground colour of the Gold variety is a bright rich bay, and the striping and spangling on the feathers is black with a rich beetle-green sheen. A Silver variety also exists, but is very seldom seen.

The OEPF is a very hardy breed. The author well remembers being told by one fancier how he dusted frost off the backs of some Pheasant Fowl that had decided to sleep out on the fence in winter and that they suffered no ill-effects as a result!

OEPF are alert, active and very reasonable laying birds. Being good foragers they are best suited to free range. They are suitable for a novice keeper with plenty of space to keep them.

These attractive birds are never found in great numbers, but deserve to find more breeders, as they make an excellent smallholder's fowl.

Left: The rear spike of the rose comb should follow the line of the neck but not touch it.

Poland

'Poland' is a bit of a misnomer for this breed, as it was not created in that country. The name possibly derives from the word 'polled' as used to describe the dome-shaped head seen in some cattle breeds. Polands do have an odd-shaped skull, something that was noted as far back as 1656.

The Poland is a good laying breed, known for centuries as a producer of large eggs. Historically, it was much used in France to cross with other breeds to increase egg size and production.

There is a Frizzled variety with curled, forward-facing feathers.

General care

Polands must be kept dry at all times, so covered runs will be needed for most of the year, except in the best of weather.

The crest of the Poland should be circular and very large, and it is this feature that presents the majority of problems encountered in the breed. The birds are very prone to suffering from mites in the crest and, in the case of the bearded variety, in that too. A crest is not something a bird can

Left: The crest of the Poland (here a silver-laced) should be very round and globular. It should not be permitted to get wet.

preen clean itself, so if it becomes soiled the owner must ensure that it is washed and dried.

Water should never be provided in open containers, as a wet crest, especially in freezing conditions, can cause the bird great suffering.

Polands should not be kept in a flock with other breeds as they can be badly bullied. Other birds will frequently take to pecking at the crests of growing Polish chicks.

Since it has a general lack of anything except forward vision, creeping up behind a Poland can give the bird a frightful shock. Always speak before picking up a bird so that it knows you are there. The Poland is totally defenceless against predators and should be protected at all times.

This breed can also suffer from eye infections and owners need to be on a constant lookout for this.

Above: The laced colours (here a Chamois) have beards; non-bearded varieties are available.

Points to bear in mind

Clearly, the Poland is a very high-maintenance breed that needs the care of an experienced and knowledgeable poultry keeper if the birds are to remain healthy.

Redcap (Derbyshire Redcap)

The Redcap developed in the North of England, alongside breeds such as the Hamburgh and Old English Pheasant Fowl. In the early days, fanciers would meet at inns and clubs to compare their fowl and compete for prizes, and the breed was already well developed by the 1860s. In a report of exhibition poultry at the fifth meeting of the Sheffield Society in 1861 we find mention "That local favourite of the Sheffield district, the "Redcap" mustered strongly, and seemed to attract much attention."

In 1867 Tegetmeier writes "The cocks not unfrequently possess combs upwards of three inches in breadth at the front, and more than four in length, measured to the end of the peak behind. Valuable as Redcaps may be, both as table fowl and as enormous egg producers, we cannot do more than regard them as a local breed, not likely ever to rise into general estimation."

It has always been very much a local breed, with most still being kept around the Derbyshire area. In recent years, with more people becoming involved in rare breeds of poultry, interest in this rather unusual breed is once again starting to blossom.

Above: *It will take up to three years for the male Derbyshire Redcap to acquire the full magnificent comb reflected in the breed name.*

Appearance and temperament

Redcaps have always been regarded as a robust, long-lived breed. They are good layers and very good foragers, preferring free range environments. The hens are non-sitters and hence do not go broody. The plumage of the hen is a most attractive nut-brown colour with neat spangling, and the striking male has orange-coloured hackles. The exceedingly large and well 'worked' comb, with plenty of fleshy but even-height spikes, give the top of the comb a flat appearance. This breed should always have red earlobes.

Points to bear in mind

If the breed is to be kept where below-zero temperatures are maintained for any length of time during the winter, it may be best to smear the comb with petroleum jelly, as a precaution against frostbite.

Scots Dumpy

Breeds of poultry that carry the short-legged 'creeper' gene have been known for centuries, and very similar breeds have emerged in different countries. The Scots Dumpy made its first recorded appearance at a poultry show in 1852. It has had different names over the years, 'Bakies', 'Creepers' and 'Crawlers' being amongst the most common. The breed was very popular in the UK until the 1870s, when numbers declined as other, newer breeds arrived. Interest grew again after the turn of the century, only to fall away rapidly. The breed became almost extinct after the Second World War. It was revived in 1973 using the few remaining birds,

together with imported stock from Kenya. The Scots Dumpy Club took over the management of the breed in 1993. Although not classed as a rare breed, it is certainly very scarce.

Appearance and temperament

The body shape of the Scots Dumpy should be long and low with a boatlike shape, giving the appearance of being very heavy. The most usual colours are Black and Cuckoo, with White making an occasional appearance.

This is a docile breed, suitable for the novice keeper. The hens are reasonable layers and do make good broodies, but dislike being moved once they have decided where they want to sit.

Points to bear in mind

The creeper gene is semi-lethal, which means that 25% of the fertile eggs will fail to hatch, resulting in fewer chicks than would occur in other breeds. The shortest-legged males can suffer from infertility problems. They do best kept away from muddy ground.

Left: *The boat shape of the body and the very short legs are typical of the Scots Dumpy. This is a Cuckoo pullet.*

Scots Grey

As the writer Edward Brown states in his book of 1929 "Whence the Scotch Grey was derived is buried in oblivion." The breed is certainly very old and was without doubt a cottager's fowl, found amongst the smallholdings of Scotland during the nineteenth century. It was described as "a first class all-round fowl" and claimed to be noted for laying an egg that required "a larger egg-cup than the Staffordshire potteries are accustomed to make."

The breed name was changed to Scots Grey in the early 1920s, as 'Scotch' actually referred to the native whisky, rather than the country.

Appearance and temperament

The colour of the Scots Grey is very precise, with a neat, clear, barring pattern of metallic black and steel grey. Unfortunately, really good-quality barring is often only found in the Bantam version, with the heaviest specimens of the Large Fowl tending towards a more indistinct pattern. The overall colour of the bird should look even from head to tail. The body shape needs to be fairly long, with a well-rounded breast.

The Scots Grey is a dual-purpose breed, hardy and easy to rear, but egg numbers are not particularly high. It does well on free range, but accepts confinement without much argument.

This very rare breed needs a lot more support than it attracts at present. It is most popular in its native Scotland, but seldom seen at more southerly shows. When it does appear it usually attracts admirers.

Left: *The Scots Grey is suitable for the novice keeper, but locating available stock may be a problem. The crisp, clean barring is very noticeable in this breed.*

Sicilian Buttercup

The Sicilian Buttercup is believed to have been originally imported into the United States from the island of Sicily in about 1835. However, it was with later importations from about 1860 that the breed became established. It was introduced into the UK from the USA around 1913, but is now very scarce.

The name Buttercup comes not only from the gold-coloured plumage, but also from the unusual shape of the bird's comb. This is the result of a mix of two distinct genes: the duplex gene that causes the horned comb, as found in many of the crested breeds, and the normal single comb gene. When combined, the result is a circular comb that in the best specimens looks like a circle or cup. It is difficult to get a Buttercup with a perfect, circular, cup-shaped comb. It is the norm for many chicks to hatch with an imperfect comb.

In Sicily it is believed that the breed originated as a cross in ancient times between an Italian fowl and an African one. The cup-shaped comb is a trait the Sicilian Buttercup shares with the Egyptian Dandarawi fowl, so it is very possible that the two are related.

Appearance and temperament

The female's rich gold coloration, with its striking black patterning, is one of the most beautiful features of the breed and attracts many admirers.

The Sicilian Buttercup is an active breed that really enjoys the chance to forage and free range. The hens are not prolific layers, but will produce a reasonable number of small white eggs.

Below: *The cup-shaped comb should have at least one normal spike at the front before dividing off to either side to form the cup – a perfect cup comb is very hard to find.*

Silkie

The explorer Marco Polo mentioned a breed with "hair like a cat" in his writings about the Orient, which provide a documented date of around 1298 for the Silkie. However, there is no doubt that the breed is far older. Its historical use in Chinese medicine is recorded, and recent studies have revealed that the flesh of the Silkie contains a far higher proportion of carnosine, a compound with antioxidant properties, than is found in normal poultry.

Appearance and temperament

The Silkie is remarkable, as it not only has black skin but dark flesh as well. All colours should have dark skin, so avoid any with red faces. The birds have five toes on each foot, feathered legs, turquoise earlobes and a cushion comb.

The feathers have no barbs to hold them together, as found in normal feathers, and this explains the fluffy, hairlike appearance of the breed. These disunited feathers also mean that this breed is not waterproof and must be rigorously protected from bad weather.

The Silkie is widely used for crossing with other breeds to produce excellent broodies. The hens' main ambition in life seems to be permanent motherhood. They make good winter layers, often continuing to lay when other breeds have called a halt. The Silkie is a very docile, gentle breed, suitable

Left: This large Blue-bearded Silkie pullet shows even plumage colour and good breed type.

for the novice prepared to give it extra attention. Because of its small size many people wrongly think that all Silkies are Bantams, but the Silkie is actually classed as a Large Fowl. A genuine Bantam version has recently been developed and this is very small indeed, weighing only around 500gm.

Points to bear in mind

Some strains are prone to a paralysing disease called Marek's, so only buy stock from an experienced breeder and talk to them about this problem to discover if the birds have been vaccinated. Keep Silkies out of the mud and rain, and check them on a regular basis for scaly leg mite.

Left: The face of the Silkie should always be dark in colour, as in this Bantam male.

Spanish

It is known that amongst breeds with white earlobes, the white colouring can drift into the face. This is obviously what happened in this breed, with the white face being selected as a point of merit. Illustrations of the Spanish breed from 1801 do not show the white faces, yet the text does make specific mention of the white earlobes. Later, in *The Mirror of Literature, Amusement, and Instruction* of 1833, we find mention of: "fine black Spanish birds as ever was seen, with combs as big as beef steaks and white ear-bags just like pillow cases".

The Spanish was certainly one of the most popular breeds in the early poultry shows from around 1850 onwards, with breeders trying to out-compete each other to produce the largest white face on the breed.

As with many breeds, popularity waned and by the mid 1920s the Spanish was almost extinct. Luckily, it was resurrected from a few remaining specimens by some rare breeds enthusiasts in the 1970s.

Appearance and temperament

The Spanish should have a long white face with as little red in it as possible and no blemishes. This is very hard to achieve. This breed cannot be recommended as a first choice for the novice.

Right: A Black Spanish female showing the unusual white face only found in this rare breed.

Like all Mediterranean breeds the Spanish is active and lively. They enjoy free ranging and produce a reasonable number of good-sized eggs.

Points to bear in mind

The Spanish remains very rare, and is a breed for the dedicated exhibitor. The birds' faces suffer from exposure to the elements, especially during freezing conditions. To keep the faces looking good takes a great deal of dedication, as they can easily become damaged and blemished. To present a Spanish with a perfect white face for show means long hours of tender care and much dusting with baby powder!

The males tend to be short lived and damage to their faces can make them very unattractive. The hens tend to peck at the male's face, which will cause scarring. Provide extra protection during very cold weather.

Sultan

When Miss Elizabeth Watts first spied her consignment of Sultan fowl on the English shores, they were a sorry sight indeed. Imported from Constantinople in 1854, the birds were very much the worse for their travels. Writing in *The Poultry Yard* c. 1870 she says "The voyage had been long and rough; and poor fowls so rolled over and glued into one mass with filth were never seen." She goes on to say that it was impossible to tell even what colour the birds were.

Amazingly, the tough little fowls survived their experience and moulted out into the wonderful white breed seen today. All present-day Sultans are descended from those few imported birds, and their name derives from that of the breed in Turkey, which was Serai-Täook, meaning 'fowls of the Sultan'.

The Sultan's path to the twentyfirst century has not been an easy one. The breed started off as very rare, even in its country of origin, and is still classified as such today. It has never been kept in large numbers, but due to the dedicated work of a few fanciers, Sultans are still available today.

Appearance and temperament

The Sultan is found only in Large Fowl in the UK. A Bantam version has been developed in

Right: Birds with horn-shaped combs always have large prominent nostrils.

the USA and coloured versions have appeared on the Continent in recent years. Although classified as a Large Fowl, Sultans are not a big breed, weighing only around 2-2.7kg.

The Sultan surely has the most ornamental features of any breed of domestic poultry. The birds have a crest, beard, five toes, a horn comb and large cavernous nostrils. They have stiff feathering on their legs and feet and a feature known as 'vulture hocks', where the quill feathers growing on the hock are stiff and point backwards. In addition, Sultans are pure white in colour.

Taking all these points into consideration, it is clear that the Sultan is not a breed to be undertaken on a casual whim. It is neither a table bird nor a laying bird, but is kept purely for its beauty and outstanding ornamental features. Sultans are calm, good-natured fowl, the hens in particular becoming very tame.

Keeping Sultans in perfect conditions in a temperate climate is an uphill task, but the dedicated owner can meet the challenge. Sultans are not best suited for free-range environments unless the weather is dry. Muddy and wet ground is a disaster for them, as their stiff foot feathers quickly become clogged with mud and then break.

The perches in their house should be no higher than 40cm from the ground, and it is a good idea to allow extra perch space in order not to crowd them. Keep their bedding very clean and at least 10cm deep to help prevent damage to the foot feathering.

Left: Always provide Sultans with water founts to drink from and not open bowls.

Right: With its white plumage and long feathered feet, keeping a Sultan clean and tidy for showing is a hard task.

Points to bear in mind

When buying Sultans, make the usual health checks, but also remember to check the toes of the bird. There should be five toes and the fifth toe should emerge from the shank, rather than just being joined to the normal rear toe. Make sure that the bird has the correct horn comb. Be aware that breeders will be very reluctant to sell single hens, as the gene pool is so small.

Because the crests restrict backward visibility and the large foot feathers restrict rapid movement, these birds are easy prey for any passing animal that wants to attack them.

Check for northern mite regularly, especially in the crests and beard, and check for scaly-leg mite often, as all feather-footed breeds seem especially prone to this pest. The Sultan's crest should not cover its eyes, but birds with larger crests can occasionally get eye infections and this is something to watch out for. If you are not showing your birds, then trimming back the crests is an option.

When breeding Sultans, it is advisable to trim back the foot feathering, as this assists the male when treading the females.

Sumatra

In the *Transactions of the Essex Agricultural Society, Massachusetts* of 1850, we find Sumatra Game listed among a group of fowl present at a show, for which S. & O. Southwick from Danvers were awarded the sum of $6 in prize money.

Originally known as the Sumatra Pheasant Game, (or Ebon Game) this name was later shortened to Sumatra Game. Its origins lay amongst some fighting fowl that were imported from Sumatra into the USA in the 1840s. The breed was soon developed away from its fighting roots and became purely an exhibition fowl. Today, it shows no more pugnacity than any other soft feather breed. Because of this, the 'Game' suffix was dropped and the breed is now referred to simply as the Sumatra. It became standardised in the UK in 1906.

Appearance and temperament

The most noticeable thing about the Sumatra is the length of tail in the male. It should be very long and carried at a low angle, similar to that of a pheasant, with the standard calling for it to hang just above the ground. The females also display a tail that is longer than that of the average hen. The Black version should have a brilliant metallic-green sheen to the feathers. The breed is also found in Blue and White, with Bantam forms available in all these colours.

In the UK the breed is standardised with very dark-coloured faces, black facial skin being preferred. In other countries it is normal to find the breed with a redder face. It should have multiple spurs, with males having anything from three to five spurs on each leg.

The Sumatra lays a white egg and the hens make excellent broodies and mothers.

Above: *The lustrous sheen on the feathers of the black variety should be green and not purple.*

Points to bear in mind

Take special care to keep this breed clean and its tail feathers unbroken. Provide high perches set away from the walls. It should go without saying that it is a disaster to allow the males to get their tails muddy! However, with a bit of extra care this breed can be kept by the novice.

Below: *The Sumatra is a fowl that prefers living on extensive range. They will often take to sleeping in trees, as they have strong wings and fly well.*

Vorwerk

In about 1902, Oskar Vorwerk set out to create a breed of rich buff-coloured fowl that also sported the solid black neck hackle found in the Lakenvelder (see page 102). He was aiming for a good-quality utility breed that would mature early and carry more flesh than many of the other farmyard breeds found in Germany at that time. He increased the size of the breed, and at the same time introduced the buff colour using crosses between Buff Orpingtons, Lakenvelders, Andalusians and Buff Ramelslohers. His breed was first exhibited in 1912.

Although it was nearly lost after the Second World War, this attractive-looking breed was revived from the remaining stock. It was imported into the UK in 1970 and the Bantam form arrived in 1997. Since then a Blue version of the Bantam has been produced.

Appearance and temperament

The Vorwerk should be similar to a Welsummer in size but, unfortunately, fowl that appear to be somewhere between the Bantam and Large Fowl in size often appear at sales. There is no excuse for breeding from inferior quality stock just because the breed is classed as 'rare' – it does the breed no favours.

The Vorwerk is a very practical breed, very suitable for the novice keeper. The birds are good layers and fairly fast to mature.

Much is often said about the compatibility of the males. In the author's experience, males raised together from an early age will get along with each other when adult. However, if they are not raised together, they are no more peaceable than any other breed. Bantam males can be aggressive and are best kept firmly away from children. However, Large Fowl males are generally very good tempered.

Points to bear in mind

It is difficult to tell how clear the top colour of a Vorwerk will be until it starts to acquire adult plumage. Few people are prepared to keep birds this long, only to find that they are not what is required.

Left: *Vorwerk male. The undercolour of the feathers should be grey and not buff.*

Right: *Vorwerk Blue Bantam female. In this variety a rich pigeon-blue colour replaces the black.*

Welsummer

This breed was created in the early years of the twentieth century in the IJssel-Valley and takes its name from the village of Welsum in the Netherlands.

It is believed that Faverolles, Wyandottes, Brahmas, Cochins, Malays and Dorkings were all used in the original makeup of the breed. Not surprisingly, it took a good deal of time to get the breed to a consistent standard. Barnevelders were added later to help add a more uniform shape. However, not content with this mix, breeders then added Rhode Island Red and Partridge Leghorns to increase the egg production!

Left: Welsummer hen. The eggs are a rich dark brown. They are often exhibited in egg classes at shows.

Appearance and temperament

The breed found today is remarkably stable in appearance, and many people refer to it as the 'perfect-looking chicken'. Although colours such as Duckwing are occasionally available, the standard Black-Red male and Partridge-coloured female remain the norm.

The Welsummer is an active, alert breed and very good at foraging. It is slightly on the large size when compared to other Light Breeds, almost certainly reflecting the various Heavy Breeds used in its original makeup. A Bantam form is available and equally popular. It is suitable for the novice keeper.

The breed can be a bit nervous, and tends not to be quite as tame as some other breeds.

Welsummers are always in great demand as they are good layers of superb terracotta-coloured eggs, often described as a 'flowerpot colour'. The strains that lay the

least eggs tend to lay the deepest-coloured ones, so you may have to compromise between egg colour and numbers if you require large quantities of eggs.

Points to bear in mind

Many strains of Welsummer produce chicks that tend to feather-pick if bored, so make sure the environment for the chicks provides plenty of alternative mental stimulation.

Left: The Welsummer is a very good choice for the novice poultry keeper and is widely kept, so there should be no problems locating stock to buy.

Yokohama (Phoenix)

Although the name indicates its Japanese origin, there is no poultry breed in Japan that goes by this name. The Yokohama appears to have been named after its port of departure for western shores. In the UK, all the sizes and colours of these Japanese long-tails are known as Yokohama, although on the Continent the name Phoenix is often applied to the single-combed variety.

Long-tailed breeds have existed in Japan for many centuries, but first arrived in the West in 1864. They were imported into France by Monsieur Girard, who had been a missionary at the cities of Yedo (modern day Tokyo) and Yokohama for a number of years.

All European long-tailed breeds are related to a breed called the Onagadori (see page 121). It carries a non-moulting gene and thus continues to grow its feathers all year round. The European versions of the long-tailed fowl do moult, but frequently not until their second year, and the tails continue to grow during this time. When first imported the breed was

not vigorous, and various European breeds were crossed in to obtain more robustness.

Appearance and temperament

The single-combed variety is very similar to many of the Japanese breeds in appearance. The body shape of the Yokohama should be long and tapering, with a pheasant-like appearance. Without doubt, the main attraction of this breed is the amazing length of the tail; even the hens have a very long tail, with the two top feathers showing a graceful curve. The Yokohama is generally a very nice-natured breed. It responds well to being handled and tames fairly easily. It certainly makes an exotic-looking pet, and fly fishermen are always keen to try using any spare tail feathers for fly-tying.

Points to bear in mind

Perches need to be high and set well away from the walls of the house to prevent the tails becoming soiled and broken. No variety of long-tailed breed should ever be permitted to drag its tail through the mud. The care of the tail requires experienced husbandry, so for this reason the breed is not suitable for the novice keeper. It is also fairly prone to respiratory infections.

Left: Note the very long tail with abundant side-hangings of the male. The breed can also have a pea or walnut comb.

Yokohama Phoenix male.

119

Yokohama Red-saddled White

The Red-saddled White Yokohama was developed by Herr Hugo de Roi in the late nineteenth century around the area of Brunswick, Germany. There is no equivalent breed type found in Japan.

Appearance and temperament

The striking coloration is unique to this breed. The neck and tail are pure white, while the body is a rich crimson in the male and red-buff in the female. In addition there is a distinct white spangling at the end of each feather. With its walnut-shaped comb, the Red-saddled White Yokohama has a far more gamelike appearance than the single-combed variety, this feature being most noticeable in the Large Fowl.

The tail feathers of all long-tailed fowl should be supple and soft, as this means they are less prone to breaking. It takes a great deal of skill and dedication to preserve the tail feathers of a male in quality condition, especially when those tail feathers are meant to stay pure white!

The Bantam form was first developed around the 1920s, but not standardised until 1968. It has now become the most popular form, with the Large Fowl very seldom seen in the UK. The gene pool of the Large Fowl is now very small, and imports from Europe will no doubt be needed before the Large Fowl Red-saddled White Yokohama becomes a spectacular sight at UK shows once again.

In temperament these are very inoffensive birds that respond well to handling. They fly well and far prefer sleeping up trees than being confined to a hen house. They are not brilliant layers, but with their exotic looks they deserve to be allowed some faults!

Left: *A walnut-shaped comb gives this striking breed a very gamelike appearance.*

Points to bear in mind

The breed is fairly prone to respiratory infections. Perches need to be tall and set away from the walls of the house to prevent the tails becoming soiled and broken. Needless to say, a Yokohama should never encounter any mud, as the care of the tail, especially if required for showing, requires exceedingly good husbandry.

A deep crimson red colour is set against the white background with white spangling on the breast.

Left: *A rare Large Fowl Red-saddled White Yokohama male.*

The Onagadori

In Japan, the Onagadori has been known for centuries and is regarded as a national treasure. It was developed to produce the long feathers required for the standards carried by the Imperial standard bearers. The early form of the breed was almost certainly the Shokoku, which is still in existence, but the development of the mutant non-moulting gene produced the Onagadori. This gene causes the birds to grow tails that can measure over 10m long, the record being 11.3m. The breed only really came to the attention of the Western World after an eleven-page article by Frank Ogasawara appeared in the December 1970 edition of National Geographic. The accompanying photos showing a male bird being taken for a walk, with its keeper following behind carrying the vast train of feathers, was a total sensation. Onagadori are occasionally seen on the Continent, but none have been seen in the UK for many years.

Introduction to Heavy Breeds

If we look back through the history of the breeds, it becomes clear that breeds classed in the UK under the heading of 'Heavy' generally all have Asiatic breeds somewhere in their background. The introduction of these very large Asian poultry breeds radically increased the weight and size of the fowl bred for the table.

The Orloff is one of the rarer Heavy Breeds.

When considering poultry breeds, remember that 'Light' and 'Heavy' refer to the Large Fowl versions of the breed. Although you may be looking at a very small Bantam, it owes the classification to the utility merits of the Large Fowl breed it represents in miniature.

The Brahma is an example of the Asiatic breeds.

Alternative terms often used to distinguish the two types of fowl are 'sitter' and 'non-sitter', which relate to the tendency of the Heavy Fowl to go broody easily and the Light Fowl to ignore the idea of broodiness and motherhood altogether.

However, rules are made to be broken, as they say, and there are some notable exceptions. For example, the Silkie – regarded as the best sitter of all and a breed of Asiatic origin – has always been classed as a Light Breed in the UK. This is almost certainly because in size, the original Silkie fitted midway between Large and Bantam, and to call it 'Heavy' would have been plainly ridiculous.

Simply put, we can say that breeds designed as table fowl or the big-boned dual-purpose breeds fall into the category of Heavy Breeds.

These birds are usually placid in nature, friendly, easy to look after and very robust. It is from this section that the novice would be well advised to seek out their first choice of breed.

Australorp

Australorps are the Australian refinement of the Orpington breed (see page 154), developed from strains of the original black Orpingtons that were exported to Australia during the 1880s.

While British breeders plunged with great enthusiasm into endowing their version of the black Orpington with a vast profusion of feathers and show merits, the Australians were busy working towards a far more practical and utility form. The resultant fowl was a breed that by 1923 could set world records for egg production. The name Australian Utility Black Orpington was soon shortened to Australorp.

Appearance and temperament

There is far more to the Australorp than a black bird with glossy plumage. Clean fine lines, a perfect-shaped head and the glossiest of black plumage, plus a good utility shape, are all points that spring to mind when looking at the Australorp. To produce a bird with exactly the correct confirmation, outline and head points is not as easy as it looks. Avoid breeding from birds with white colour in the earlobes, as this is a serious fault when showing the breed.

Sadly, Australorps suffer from the 'black bird unpopularity syndrome'. Many people are inclined to pass over breeds with black plumage, thinking the colour boring, but do take a closer look at this breed, which really does have a great deal in its favour. That beetle-green gloss on the feathers is quite spectacular when viewed on a sunny day.

A Black Bantam variety was developed in the UK prior to the Second World War, and both Blue and White versions of the breed are now available. The Bantam form now far outnumbers the Large Fowl at poultry shows.

This is a very practical breed; the birds make good broodies and mothers, and are very decent layers, producing around 200 eggs a year. Being a heavy breed, spare cockerels are good meat birds. In general, they have a very even, docile temperament.

This breed should offer the novice no obvious problems and would be a very good choice.

Left: Alert, very dark eyes and a wonderful green sheen on the black variety are just two of the Australorp's qualities, along with a good laying ability.

Barnevelder

This is a Utility breed, developed in the area of Barneveld in the Netherlands. It was imported into the UK in about 1921 and attracted immediate interest due to the fact that it was both a good winter layer and a producer of brown eggs.

Being a very robust breed, the Barnevelder should pose no obvious problems for the novice keeper. It would prove an excellent choice, combining an attractive feather pattern with good utility features.

Appearance and temperament

Originally, breeds such as Brahma, Cochin and Langshan were used in its makeup, but despite all these feather-legged ancestors, the Barnevelder has clean legs that are yellow in colour.

The most popular Barnevelder is the very pretty Double-laced, which has two rings of glossy greenish black lacing on a nut-brown background. A Blue-laced version has since been developed and it is also standardised in Black, Partridge and Silver, although these latter colours are rarely encountered.

The Barnevelder is a very solid, reliable and dependable breed. The males are usually good natured, but as in all breeds exceptions can be found during the breeding season. Excess males make good table birds.

Eggs

Barnevelders lay a good-sized brown egg. It is possible to find strains that lay pale-coloured eggs; avoid these if possible, as one of the key factors of the Barnevelder breed is the brown egg. Avoid breeding from birds that produce a pale egg colour.

The Bantam form is very well established, both as utility Bantams and as show birds, but the egg colour can be paler than that found in the Large Fowl, although again some strains are better than others.

The Barnevelder makes a good practical breed for the novice.

Left: *A Barnevelder Double-laced female. The plumage is very attractive, especially when well laced. The best Barnevelders are easily capable of taking top awards at shows.*

Bielefelder

This relatively new breed has already become very popular on the Continent and some Bielefelder have now made an appearance in the UK. Created during the 1970s by Herr Roth of Bielefeld in Germany, it has the advantage of being both an autosexing breed and a layer of very dark brown eggs.

Unlike most autosexing breeds, the Bielefelder was developed quickly into both the Large and the Bantam form. They are already proving popular in both sizes, not only in Germany but also in other European countries.

The breed was created using the Dutch breed the Welsummer and the Cuckoo Malines, an old Belgian breed originally used for table purposes. This parentage has resulted in a breed that is quite a bit heavier in type than the existing UK autosexing Welbar breed (see page 89). Males weigh about 4kg.

Appearance and temperament

The males are a very eye-catching mix of Cuckoo and Gold, with bright yellow legs. The hens are similar in colour, but with a much softer marbling effect. The hens are said to be good layers, with early-hatched pullets laying well throughout the winter months. The breed is calm-natured, tames easily and is vigorous and hardy. It is said to tolerate cold weather well.

Left: This breed has an attractive plumage colour. It should give no problems, although locating stock could prove difficult.

Brahma

The Brahma is a huge and impressive bird, with males weighing about 4.5-5.5 kg. It is often referred to as the gentle giant of the poultry world, and its placid good nature and easy-going ways win many a convert to the world of poultry keeping. These gentle birds make excellent pets for children.

Origin

The history of the breed is fascinating, as there are a number of different versions of its origins depending on which history book you read.

In 1853, George P. Burnham, an American entrepreneur, author and self-proclaimed publicist, sent a pen of these birds to Queen Victoria in England, in an effort to receive as much publicity as possible for his own stock. His ploy worked and a 'Hen Fever' was kindled, with people paying vast prices for the birds. Although previously known as Grey Chittagong or Shanghai, Burnham claimed he named the breed Brahma Pootra after a River in India. This name was later shortened to Brahma, although Burnham himself humorously liked to call the breed the Bother'em Pootrums.

The breed itself was actually imported from Northern China

and has a close relationship with both the Cochin and the Langshan. It was the British and American breeders who refined the breed into the stunning-looking fowl we have today.

Appearance

The Brahma is found in various colours. The Light is a Columbian pattern, similar to a Light Sussex. In the Dark, the females have a delicate pencilling pattern, with fine black lines on each feather. In the Buff Columbian, an even buff shade replaces the white in the Columbian pattern. Female Gold Brahmas have a rich gold background, with black pencilling similar to the Dark variety. Other colours are now being developed, such as Blue Partridge, Pile and Blue Columbian. Look out for a small pea comb and overhanging eyebrows. Above all, Brahmas should be really big and impressive. The breed is not particularly active, far preferring to strut around and look stunning. It is very suitable for the novice, with a bit of extra care.

Brahma chicks can be sexed at an early age due to a sexlinked slow feathering gene, which means that the males remain very sparsely feathered for a number of weeks.

Brahmas are very slow growers and not really suitable as a meat breed. Their eggs are neither

Left: Brahmas are known for their 'beetle brows' that can give them a rather stern expression. Note the intricate lacing on this dark female.

Points to bear in mind

When breeding Brahmas, you may need to trim away the feathers around the vent area to aid fertility.

As with all feather-footed breeds, check regularly to make sure they are not suffering from scaly leg. It is a good idea to have somewhere suitable for them to go when the weather is wet, as they do not do well in muddy conditions.

The Blue Partridge coloration is a recent addition to the breed.

Above: Dark Brahma male. The Brahma is a wonderful breed and can be highly recommended to those who want something stunning, placid and gentle to admire.

very plentiful nor large, which can be very disappointing for those expecting something substantial from this massive bird.

Housing

Brahma houses will need a little adaptation. Perches should be low to avoid damage to the birds' legs and feet when they jump down in the morning, and doors must be big enough to accommodate their huge frame.

Brahmas are not escape artists and will stay behind a low fence with no problems, as they do not fly. Protect them from predators, as they cannot quickly escape from marauding foxes and dogs.

127

Buckeye

This breed was produced by Nettie Metcalf of Warren, Ohio, in 1896. Writing in 1907 of their origin, T. F. McGraw and George E. Howard said: "The Buckeye was originated by Mrs Metcalf of Warren, Ohio and was originally called Buckeye Red. In the production of these birds a cross was made of the Asiatics, Black Red and Indian Game."

Although Mrs. Metcalf made no mention of using Indian game in her earlier writings, she later said that she felt the Black-Red gamefowl that she used had been part Indian Game, as some of them had pea-combs and yellow legs. Certainly the breed does carry a strong resemblance to the early forms of the Indian Game fowl.

The breed identity was nearly lost when Mrs. Metcalf was almost persuaded by a breeder of Rhode Island Reds to rename her breed as a pea-combed Rhode Island Red. However, just in time she realised that the type of fowl she was aiming for was of a far stockier type, so she declined to do this. She named her fowl Buckeye after the colour of the Buckeye nut found in Ohio. This is much like the colour of a horse chestnut. The breed was admitted to the standards in 1904, the same year that the Rhode Island Red was admitted.

Appearance and temperament

Mrs. Metcalf specifically wanted the plumage of her breed to show some dark pigment in the underfluff, as she firmly believed this gave better colour to the surface feathers. Hence, the underfluff of this breed is red, except on the back which should show a bar of slate colour when the feathers are parted.

The Buckeye is a tough, robust dual-purpose breed. It has never been hugely popular and is still classed as rare, even in its native country.

Points to bear in mind

Locating available stock can prove a problem, as the breed is still scarce in the UK. Some males are reportedly rather aggressive during the breeding season, so it would be best to keep them away from children until you have assessed their temperament.

Left: Parting the feathers on the back will show a bar of slate colouring in the underfluff. A Buckeye should look very powerful.

Chantecler

This is a breed created out of patriotism. In 1908 The Reverend Brother M. Wilfrid Châtelain decided that Canada should not be without its very own breed of poultry. Therefore he embarked on producing a breed that would be not only a very good table fowl, but also a good winter layer, strong and hardy. He had seen for himself that large-combed birds tended to suffer from frostbite during the cold Canadian winters, so wanted his breed to have a very small comb. In his own words he wanted a breed "without any fantastic or whimsical features".

Initially, he made use of Indian Game (Cornish), White Leghorn, Rhode Island Red and White Wyandotte. Later, he would introduce White Plymouth Rock. It was to prove a long and very persistent breeding programme, using various

Below: *The breed was designed to withstand the fierce Canadian winters. Therefore they have small, close-fitting combs and small wattles.*

Above: *It is estimated that there are around 2,000 Chantecler in existence today. Most of these are still located in Canada. This is the original white form.*

crosses and back-crossing of birds. All his work was carefully documented. Eventually he had the fowl he had set out to produce and the vigorous breed of the Canadian Chantecler was admitted to the American Poultry Association standards in 1921.

In 1927 Brother Wilfrid submitted a paper on how the breed was created to the World Poultry Congress in Ottawa, describing in detail all the various crosses he had used to create this truly Canadian breed. He was later awarded a Doctorate in Agricultural Sciences.

Appearance and temperament

Today the Chantecler is found not only in the original White form, but also in a Partridge colour. A Bantam form was also created during the 1960s. Sadly, the breed is very rare and tends not to be found outside Canada or the USA. However, as many rare breeds now seem to be arriving in the UK from overseas, it may be only a matter of time before the Chantecler makes an appearance.

Cochin

This is the second of the large Asiatic breeds that caused the 'Hen Fever' of the 1850s (see page 126).

In George P. Burnham's book *The History of The Hen Fever* written in 1855, we get an idea that the breed was not exactly standardised when it first started to become popular. Burnham states "Their plumage was either spotted and speckled or it wasn't. And thus the true article, the pure-bred Cochins, could always be designated and identified, – by the knowing ones I presume. I studied them pretty carefully, however for five years; but I never knew what a "Cochin-China" fowl really was, yet!".

Thankfully, this docile breed did get refined to a high standard a few years later. Birds found today should meet it if they are to be used for breeding or showing.

Below: *This Buff Cochin hen shows the sort of width required in a show specimen.*

Right: *Cochins should display a matronly appearance, with a full breast.*

Appearance and temperament

The Cochin is a large breed and should weigh in the region of 4kg. It has a wealth of fluffy feathering, and a very flat and broad back. The breast should be low and full. Females are described as having a 'matronly' appearance. Cochin are standardised in Black, Buff, Blue, Cuckoo, Partridge and White colouring.

With their heavily feathered feet, they really must be kept away from muddy ground or they will become miserable.

This much-loved breed is very gentle and well behaved. Given its excellent temperament and placid attitude, the Cochin is ideal for the careful novice, providing it is not permitted to get muddy.

Points to bear in mind

Check the wing feathers, as these sometimes become twisted in this breed. Avoid breeding from such birds.

Note that the hens will try to spend most of their time being broody, but because of their size they sit too heavily on the eggs, so this is best avoided.

When breeding Cochins, you may need to trim away the feathers around the vent area as an aid to fertility. Do keep a close watch out for scaly leg and mites. Large houses, large pop holes and low perches are essential for this breed.

Crèvecoeur

In the 1870s the Crèvecœur was much esteemed in its native France for its ability to lay large eggs, fatten easily and provide the "finest fowls of the Paris market". It was regarded as fast-maturing, carrying a considerable amount of flesh compared to bone, and was popular in the latter part of the nineteenth century as a table bird in France. However, it never really took hold in the UK.

Some poultry historians think that the breed was the result of a cross between the Poland (see page 107) and fowl that were local to the Normandy district of France. Other authors claim that it is derived directly and solely from the Poland fowl. The truth is that there is rather a mystery surrounding the origins of this fowl.

Certainly there is much about the Crèvecœur to suggest that it was not entirely derived from the Poland, as it is far larger, stockier and fuller-framed than this origin would suggest. However, there is little doubt that the Poland was involved somewhere in its makeup. What is known, is that the breed is French and one of the earliest documented recordings of it was in *The Poultry Book* by Wingfield and Johnson in 1853.

From its days as a first-class utility bird, this breed is now only regarded as an unusual rare breed, preserved by those interested in maintaining old and unique breeds.

Appearance and temperament

The Crèvecœur has a crest, a horned comb and a beard. It should be a large bird of stocky build and is usually black in colour. The breed should be a great deal larger than the Poland, weighing around 4kg. However, in reality, the birds found today tend towards a far smaller size than the standards demand. A Bantam form is available in Europe.

These birds are still layers of good-sized eggs. Being very restricted in vision, they tend to be non-aggressive and placid.

Points to bear in mind

Having both a crest and a beard can make the breed rather prone to mites. Eye infections can be a problem. Never provide open water bowls for this breed, as the crest can become wet, which may cause infections. Locating stock could be very difficult.

Left: Only breed enthusiasts tend to keep the Crèvecoeur, with the result that it is one of the rarest of the Heavy Breeds.

Delaware

This totally American breed has yet to make an appearance in the UK, but has been standardised in the USA since 1952.

The Delaware was originally developed in the 1940s from a pale colour pattern that occasionally arose when New Hampshire Reds were crossed with Barred Rocks, something that was widely done to produce fast-maturing broiler breeds.

George Ellis of Delaware refined these pale-coloured birds and created a breed that he first named Indian Rivers, later changing this to reflect the State where the breed originated. As well as making a good table bird, the Delaware also turned out to be a good backyard breed, quick to mature, with a calm and friendly disposition, and a layer of brown eggs.

Appearance and temperament

The Delaware is a very interesting colour, being mostly white but with a barred neck hackle and tail and also with barring in the wings. In the USA it very much took the place that was held by the Light Sussex in the UK, as the breed could be crossed with both Rhode Island Reds and New Hampshire Reds to produce chicks that could be sexed on hatching by their down colour.

For a while, the Delaware held an important place among the commercial breeds of American poultry, but was soon superseded for table purposes by the White Cornish x Rock hybrids that came to dominate the market.

Over recent years the breed has been quietly slipping into oblivion and was recently listed on the American Livestock Breed Conservancy list as 'critical', meaning that fewer than 500 birds exist worldwide. In response, a Delaware Club was formed to help promote the breed.

Dominique

Confusion about the early history of this American breed can be attributed to the fact that 'Dominique' was frequently used to describe a bird with Cuckoo or Barred coloration, no matter what its actual breed.

John Bennett in *The Poultry Book* published in Boston, USA in 1851, claimed not to have seen any variation in the breed over the previous 30 years, describing them as a "very perfect breed". This places the Dominique as a distinct and unchanging breed from about 1820.

In *Miners Domestic Poultry* of 1853, Miner refers to the breed as the Native Dominique, being careful to distinguish it from others that are just barred in colouring. He says "It will hardly be necessary to give a detailed description of this breed so well known in every farmyard in the

The male will usually be a lighter shade than the female.

Appearance and temperament

The Dominique is a robust, hardy and very practical breed, with attractive, precisely barred plumage and an active, alert manner. It makes a useful utility bird and well deserves more support to promote it back to its rightful place as one of the oldest documented breeds of domestic poultry. Instead, it remains confined to a few dedicated enthusiasts and is a very scarce and rare fowl indeed.

Points to bear in mind

Locating available stock may prove a problem, but otherwise this hardy and practical breed should cause no major problems, even for novices. Avoid buying birds with red or yellow in any part of their plumage.

Left: *The barring of the Dominique should consist of shades of dark and light, but should stop short of being true black and white.*

country. They are a blue speckled, variegated or pencilled fowl of medium size, hardy, well formed and prolific and may be considered one of our best native breeds".

The Dominique always has a rose comb.

In 1870, the Dominique made its first appearance in the UK, but then faded from the scene as the barred Plymouth Rock became popular. Although included in the early standards of perfection, it was not until the 1980s that the Dominique again started to appear at shows in the UK, but by that time there were very few left in their native America. More stock has recently been imported to the UK and numbers are growing slightly.

Dorking

It is likely that these fowl arrived in Britain with the invasion of Julius Caesar in 55 B.C., as it is recorded that five-toed breeds were recommended by the Romans as the finest type of table fowl. The Roman writer Columella mentioned chickens with five-clawed feet in his *De Re Rustica*, written around 50 A.D. It is reasonable to conclude that the Romans would have brought fowls of this type with them, and to deduce that the Dorking was descended from these birds.

In 1787, *The Gentleman's Magazine*, writing about the town of Dorking, said "The poultry of this neighbourhood have been long famous, and great numbers are brought on market-days, and carried by higglers to London. There is a breed of fowl hereabouts which perhaps is peculiar to this country; the colour is either white or like a partridge, but there are five claws on each foot".

Whatever its original form, the Dorking has been lauded over the centuries as the finest of British breeds for the table. It became a very popular breed in Ireland, and the author James Nolan reported in 1850 that even the everyday roadside Dorking fowl owned by the Irish peasants weighed around 7-9lb (3-4kg).

Appearance and temperament

Few poultry look better than a huge Dorking in the prime of life in perfect feather on a sunny day. Conversely, few things look sadder than a poor specimen in the middle of the moult

Right: An example of the rarer rose comb variety of Dark Dorking. This cock bird displays a large rose comb only found in certain colours.

standing in the rain! The lesson is to choose your stock well, and from a reputable breeder, as you need Dorkings that have been kept pure to their own colour.

The Dorking should be a massive, heavy bird, long in the back and low to the ground. The most popular colours are the Silver Grey, the Dark and the Red. The Whites and the Cuckoos are very scarce today. The comb can be either rose or single, depending on the colour of the breed. The Cuckoo and White colours always sport the rose variety, while the Silver Grey and Red always have the single type. The Dark can have either type of comb.

When buying Dorkings, look for the right comb type to match the colour of the breed, and the fifth toe on each foot. This should be separate and not joined onto the other hind toe. The birds should be low on the leg, with not too much air between the keel and the ground.

A Bantam form of Dorking is available but this is often rather high on the leg.

The hens are very placid. Males in general are even tempered, but the odd bad-tempered one has been known.

Below: *The Dark Dorking, here a male, can have either a single comb or a rose comb. The latter is more scarce.*

Below: *The Red Dorking, here a female, is one of the rarer colours. It always has a single comb. Your arms may ache after carrying one of these birds any distance!*

Points to bear in mind

Dorkings can be prone to respiratory infections, so keep a watch for these. Houses will need large pop holes and low perches.

The Dorking is only a moderate layer, eggs never being the main purpose of this breed. However, as certain celebrity chefs have taken to promoting this breed, it may once again start to take its place amongst smallholders as a table breed. It certainly deserves a place in many more back gardens than currently house it and is suitable for the novice poultry keeper.

Do not expect to be able to order half a dozen pullets. Breeders will be reluctant to sell individual pullets, as available stock of these birds is very small. They would rather sell pairs or trios, usually preferring purchasers who wish to continue the future of this minority breed.

Faverolles

Just as the name Marans is always spelt with an 's' at the end, so the name Faverolles is also both singular and plural. This breed was developed in Northern France during the late 1860s and early 1870s. It took its name from the village of Faverolles in the Eure-Et-Loir Department.

It is believed that local chickens, such as the already well-established Houdan, were crossed with Dorking, Brahma, Cochin and possibly Langshan. From this makeup, it is plain to see that it was designed as a meat breed. However, the Faverolles also has good utility qualities, being a reasonable layer of brown eggs.

Appearance and temperament

It is easy to see why this bird was also given the rather affectionate title of 'tête de hibou', meaning head of an owl. The Faverolles sports a large wealth of beard and muffling around the face, but unlike most bearded breeds, it has a single comb. It has five toes on each foot, and feathered legs and feet, but is not as heavily feathered as the Brahma or Cochin.

Above: This hen shows the delicate colour most frequently encountered in this breed, which is called Salmon.

The Faverolles should be a massive, heavy breed. In body shape it closely resembles the pictures of the early Brahmas that were imported during the Victorian era. An adult male should weigh between 4 and 5kg. The Bantam variety is often rather oversized and frequently mistaken by novices for a Large Fowl breed.

Points to bear in mind

The Faverolles is a placid breed that should not cause too many problems, but you need to check regularly to ensure that mites do not take hold in the beard and muffling. As with all feather-legged breeds, keep a close watch for signs of scaly leg. Do not keep the birds on muddy ground. Provide low perches for these large birds to prevent leg and foot problems, and enlarge house doors to accommodate the breed.

Above: Look out for the wonderful beard and muffling around the face, and the five toes.

Frizzle

Although the Frizzle gene is recognised in a variety of poultry breeds, there is also a breed in the UK that goes exclusively by this name.

The gene that causes the feathers to curl the wrong way has been known for centuries, and frequently birds with this feather trait were referred to as "Friesland fowl" (not to be confused with the Friesian). Illustrations of birds with frizzled feathers occur in Aldrovandi's *Book of Poultry* published in 1600. Whereas today we may consider this trait cute and fluffy, one Victorian poultry writer declared that a fowl with frizzled feathers looked "as if it had been drawn through a furze bush backwards".

It is important to realise that a double dose of the gene for frizzling can result in narrow brittle feathers. Birds with this trait are sometimes humorously referred to as Frazzels, as they can look rather like a comic-book character that has been hit by a lightning bolt. To avoid producing such individuals, breeders frequently outcross to normal, smooth-feathered hens of the correct type. A few smooth-feathered birds tend to hatch each year along with the frizzle chicks. Doing this crossing keeps the feathers broad and of good quality.

Appearance and temperament

Frizzles are good natured, make excellent broodies and mothers and are often used as foster mothers for other breeds. They are reasonable layers, considering that the breed is purely designed as an exhibition bird. They are certainly a talking point!

Frizzles are found in a variety of colours, but usually only encountered in the Bantam form. Although Large Frizzles occasionally make an appearance, they soon seem to disappear from the scene again. This may be due to the fact that they need plenty of protection from the weather. Frizzle feathering is not waterproof and Bantams are easier to keep indoors than Large Fowl.

Points to bear in mind

The breed must be protected from the weather, as their plumage is neither water- nor windproof. Provide extra cover in cold weather, as the birds are unable to fluff up their feathers to trap warm air.

Below: Every feather on the body of a Frizzle should curl towards the head. Frizzled feathers can also be found in breeds such as Poland and Pekin.

German Langshan

This breed originated as a descendant from the Croad Langshan Fowl that were imported into the UK from China by Major F. Croad in 1872. In 1879, Langshans were exported from the UK into Germany and it was there that this tall, smooth-legged variety was created. A judicial addition of other breeds, such as Plymouth Rocks and Black Minorca, reduced the leg feathering and by 1904 the form that we know today was the official German standard.

Although the Large Fowl form is found on the Continent, it appears not to be available in the UK

Below: This breed should look well balanced. Try to look at the bird in profile so that you can see the wine-glass shape. German Langshans should pose no problems for the novice.

at the current time. However, the Bantam form is going from strength to strength. A specialist breed club was formed in 2008 and it no longer falls under the remit of the Rare Poultry Society. There should be little problem in finding stock, although expect show-quality stock to fetch high prices.

Appearance and temperament

The body of the German Langshan should look broad and strong, and when viewed from the top, equally broad from front to back. It is mostly found in Black, Blue and White, but other colours sometimes make an appearance at shows. Blacks should show a very glossy green sheen to the plumage.

This breed is frequently referred to as having a wine-glass shape on account of its outline when viewed from the side. The head and tail should be equal, or nearly equal, in height and the long legs form the 'stem' of the glass.

Although the birds may look a bit extreme at first glance, the German Langshan is actually a very practical breed. It lays well and is robust and hardy, so makes a good first choice for the novice poultry keeper.

Houdan

Originating in the Seine-et-Oise Department in France, the breed takes its name from the town of Houdan. It was originally developed as a dual-purpose breed, useful for both eggs and meat. Being both crested and bearded, with attractive mottled plumage, the Houdan rapidly became known as an exhibition breed once it was imported into the UK around 1850.

The Houdan was always well regarded as a layer of large eggs, but not everyone was enchanted by the breed's appearance. In a report from *Harpers New Monthly Magazine* of 1873, Madam De Linas says that she regarded the hens as "Resembling nothing so much as a tarred ball rolled in feathers, her eyes are quite hidden, and she can only see the ground directly in front of her.

Left: The Houdan requires a good deal of care to keep its fancy plumage in good order and the crest dry. It can also suffer from eye infections, which can be rather tricky to deal with.

Right: The Houdan is one of the breeds that has five toes on each foot.

Victim of her beauty she knows nothing of what is going on about her and shrieks with fright at the slightest noise".

The breed has always remained on the fringes of the poultry fancy, its fortunes fluctuating according to current trends. Although a Bantam form does exist, it is rarely seen.

Appearance and temperament

The Houdan is one of the breeds with five toes, like the Dorking. Unusually, the black mottling should also be found on the birds legs and is a feature of the breed. The crest should be large and rounded, but it should not obscure the sight, except from behind. A notable feature is the very unusual leaf-shaped comb found only in this breed.

The breed can be rather nervous. Because of its limited rearward vision the Houdan needs to be protected from predators as it is unable to detect them approaching from behind. Although the Houdan may look rather exotic, it still enjoys the chance to free range and be a normal chicken.

Points to bear in mind

Keep a close lookout for mites, as the breed can suffer from them in both the crest and beard. Eye infections can also occur if the crest is too large. To prevent the crest and beard becoming soiled, Houdans should always have water founts to drink from and never open bowls. Locating quality stock could prove problematic.

Java

When describing well-known breeds, many a poultry book mentions in passing that the breed had 'Java Fowl' in its original makeup. Unfortunately, very few then go on to describe any history of the Java Fowl itself. But as the Java was used in the creation of such well-known breeds as the Rhode Island Red, Plymouth Rock and Jersey Giant, it seems rather churlish to leave it out.

In the 1857 book *A Treatise on the History and Management of Ornamental and Domestic Poultry* by Edmund Dixon and J. J. Kerr, we find the following reference "Long Island, the headquarters of this

Below: Black Java hen. This is a suitable breed for novices if stock can be located. It is starting to make an appearance at UK shows.

The Java is only found in a single-combed variety.

variety abounds in a stout black fowl, single serrated comb and full wattles, I presume that, until we find a bird answering to Willoughby's description we must be content to call our large black fowl, Javas". We also learn that the slightest trace of a topknot is not tolerated and that the breed has good practical qualities.

The Java is one of the oldest breeds of American poultry, almost certainly derived from imported Asiatic strains crossed with local fowl. It was certainly a very popular dual-purpose bird in the New York and New Jersey areas in about 1850.

By the 1880s the fortunes of the Java had changed, and this once well-regarded breed was replaced by the very breeds it had helped to create.

Appearance

Three colours of Java are still in existence: the Black, the Mottled and the White. The Mottled is possibly the most frequently encountered. The Java is a heavy utility breed of fowl that lays large eggs. It was standardised in America in 1883.

Points to bear in mind

The only possible problem you may encounter is finding available stock, as the breed is still exceedingly rare. Although numbers are said to be increasing slightly in the USA, it is still very much endangered.

Left: A Mottled Java male, showing the breed's attractive black-and-white coloration.

Below: This Black Java male shows the long broad back required of the breed.

The dual-purpose characteristics of this breed have always been an important feature of its makeup.

Below: The breast of this Mottled Java female is full and broad.

Jersey Giant

This breed was developed in the American State of New Jersey in the years between 1870 and 1890 under the name of the Jersey Black Giant. The name derived not from the colour, but from the creators, John and Thomas Black. The Java was one of the breeds used in the makeup of the Jersey Giant, along with the Langshan, the Brahma and the Indian (Cornish) Game. It is believed that later the Black Orpington was also used. The idea was to provide a breed of poultry larger and heavier than had previously existed and able to replace the turkey as a table fowl.

It took a good many years before the Jersey Giant was accepted into the American Standards, but this was achieved in 1922.

Above: Jersey Giant chicks. It will take many months and plenty of food before they attain their full size and weight.

Left: The Jersey Giant should have a full, broad and deep breast, carried well forward.

Breeders of the time recorded show birds weighing over 7.3 kg. Today's weights are slightly more modest, but adult males should still weigh close to 6kg.

Understandably, the birds require a great deal of food and time to acquire their optimum size. They tend to build up their frame first and then fill out with flesh later. This is not a breed to rush to adulthood.

A Jersey Giant hen. The undercolour of the feathers of the Black should be slate or grey in colour.

Appearance and temperament

As well as the original Black, the breed is now found in both White and a Blue-Laced coloration. Look for willow (a greenish colour) legs on all colour varieties. On young Black birds this will mostly show up as yellow on the soles of the feet. Adult Black birds should have a green sheen to the feathers.

The breed is docile and generally well mannered. It is strong, hardy and surprisingly active for such a large breed, always happiest when permitted free access to forage for itself. The hens are reasonable layers and known for laying during the winter months.

Points to bear in mind
You will need to provide large pop holes and low perches for this breed. Expect to suffer aching arms if you carry the birds any distance!

Below: The eye colour in all varieties of this breed should be brown or black.

Below: Being strong and hardy the Jersey Giant is suitable for the novice, although stock is often in short supply.

Ixworth

Taking its name from the village of Ixworth, Suffolk, UK, this breed has a very well documented history. It is a deliberately created breed and the brainchild of Reginald Appleyard, more widely known for creating the Appleyard duck. In 1931 he set out to produce a perfect dual-purpose breed of poultry.

This breed is only found in one colour and that is White. White Orpington, White Minorca, Jubilee Indian Game, White Sussex and White Old English Game were used in its development. The combination of these breeds produced an excellent table bird and a better egg producer than many of the other dual-purpose breeds available at the time. Ixworth breeders claim that this is still true today.

The Ixworth was accepted into the British Poultry Standards in 1939. Unfortunately, the war years put a halt to most poultry breeding enterprises due to strict feed rationing. By the time peace arrived in 1945, the breed was already superseded by the commercial Broiler Breeds arriving from America. Snatched from the brink of extinction in the 1970s by some enthusiasts from the Rare Poultry Society, the breed still survives in small numbers. Because it is rather dull, with white plumage and no fancy points, it tends to be ignored by potential breeders.

Left: The Ixworth gets its strong and powerful-looking head from its Indian Game ancestors.

Right: Look for an all-white bird that looks like it would provide a substantial meal! The Ixworth is only found in white..

Although still mentioned in the poultry standards, the Bantam form has not been seen at shows for a very long time.

Appearance

It is always said that it should be possible to judge a good Ixworth in the dark or with your eyes closed, just by handling the bird for its table properties. Hence, the standard allows 40 out of 100 points for table merits. When handled, the breed is a lot heavier than it first appears.

Points to bear in mind

The Ixworth is still rare, even though it has a lot of merit as a backyard fowl. Locating available stock may prove difficult, but this breed is suitable for the novice keeper. Sexing the chicks at an early age can prove very difficult.

La Flèche

The La Flèche is a fascinating breed of French origin, said to have both Black Spanish and Crèvecœur in its ancestry. It was developed around La Sarthe in the nineteenth century, but reports of similar birds are found as early as the sixteenth century.

In 1873 a mention in the *Harpers New Monthly Magazine* claimed that the breed had a proud and rather aristocratic carriage, was a voracious eater who was easily fattened but slow growing, being not fastidious about its food.

The La Flèche never gained popularity in the English poultry markets of the nineteenth century, as black legs do not look very appealing on a platter. However, the breed was exceedingly well regarded in the Paris markets.

Appearance and temperament

Although smaller horn (or duplex) combs are found in crested breeds, it is in the La Flèche that it has been developed into the most prominent feature of

Below: *An unusual horn comb and huge nostrils are distinctive features of La Flèche. This is a male bird. The breed is still scarce in the UK and falls under the protective umbrella of the Rare Poultry Society.*

Above: *A La Flèche hen, a breed needing more support to ensure its survival.*

the breed. Combined with its glossy black plumage, it is understandable why this breed has often been given the nickname of Devil Bird.

In France the breed sports a small crest, but this is not evident in La Flèche found in other countries.

Although other colours now exist, Black is still the most prevalent. The breed is rare, and inbreeding amongst stock has caused size problems, with the result that the breed is now much smaller than it was originally. The birds' appearance is not to everyone's taste, but they do make a talking point with visitors.

La Flèche lay a reasonable number of large white eggs, but are not noted either for broodiness or being good mothers. In temperament they can be slightly wary, but respond well to regular handling.

Points to bear in mind

The main problem will be locating suitable stock. Otherwise, this unusual-looking breed should provide few problems, although birds will take to roosting in trees if the opportunity presents itself.

Langshan – the Croad

The Langshan has a well-documented history. This large Asiatic breed was imported into the UK from the Langshan region of Northern China by Major F. Croad in 1872. His niece became a staunch supporter and defender of the breed, trying to maintain it close to that of the original importations. Meanwhile, other breeders set about developing a tighter-feathered exhibition variety that eventually became known as the Modern Langshan. A number of very unfriendly public rows took place as to which was the best type. The British Croad Langshan club was founded in 1904.

Appearance and temperament

The Croad Langshan is generally a docile, well-mannered bird, although more active than other Asiatic feather-legged breeds, such as Brahmas or Cochins. The glossy black plumage with its beetle-

Above: This large male Croad Langshan exemplifies the shape required of the breed. The legs are not as heavily feathered as those of the Cochin.

The Modern Langshan

The Modern Langshan was developed around the 1890s. It has a far more gamelike appearance, with very long, lightly feathered legs and tight feathering. It has always been regarded as an exhibition fowl, and not expected to be either an egg or meat producer. It has been classed as a rare breed for many years and is very seldom seen, being restricted to the ownership of few dedicated people who enjoy keeping this unique breed.

The Australian, or Chinese, Langshan is a much smaller bird, developed with shorter legs than the Modern Langshan, but with a body shape similar to the German variety (see page 138).

green sheen is very striking, but the breed is also found occasionally in White. The Croad Langshan should look well balanced, with only a medium amount of feathering on the outside of the legs. The soles of the feet should always be pink, with no trace of yellow showing.

The Croad Langshan and the American Langshan bear a close resemblance; both are 'flashy' birds that attract attention. The USA version is taller and has a larger fanned tail. The feathers of the male's tail reach up to 40cm long.

The Croad Langshan should lay an egg that is a rich plum colour.

Points to bear in mind

Keep a close lookout for scaly leg mites. Otherwise, the Langshan should pose no real problems.

Lincolnshire Buff

The answer to the question 'When was a Buff Orpington not a Buff Orpington?' is very likely 'When it was a Lincolnshire Buff'.

Many poultry historians are convinced that the existing Lincolnshire Buff, mostly found in the Spalding area of Lincolnshire in about the 1850s, and a mix of Dorking and Buff Cochin, were the basis for Buff Orpingtons, rather than the deliberate breeding mix of Hamburgh, Dorking and Buff Cochin that William Cook claimed (see page 154).

However, the popularity of the Orpington, promoted by William Cook, soon eclipsed the Lincolnshire Buff and the breed became extinct some time during the 1920s.

Thanks to both the Riseholme Agricultural College and the dedication of breeder Brian Sands, the breed was recreated in the 1980s, finally being given an official standard by the Poultry Club of Great Britain in 1997 — nearly 150 years after it had first graced the poultry farms of Lincolnshire.

Appearance and temperament

The Lincolnshire Buff is a Large Fowl, with no Bantam form available yet. Unlike the Buff Orpington, it should not show an excess of fluffy feathering.

The birds have five toes on each foot. The fifth toe should be above and separate from the fourth toe, not joined to it. However,

as numbers of this breed are very low, it can be difficult to find birds that display really good toes.

The Lincolnshire Buff is a hardy and practical breed. Compared to many of the heavier type breeds, egg production is good at around 160 eggs in the first laying year, and the hens make good broodies.

Points to bear in mind

The breed is still very rare and locating available stock can be difficult. Breeders may be reluctant to sell to people who do not wish to breed from these birds.

Below: *A Lincolnshire Buff female. The breed is not as fluffy as the Buff Orpington and has the addition of a fifth toe.*

Marans

Due to the wonderful dark brown eggs produced, this breed (note that the name Marans should always be spelt in the plural) is one of the most popular breeds found in the UK today. It was created using a combination of the Langshan and the existing breeds in the Marans area of Western France. These local breeds contained a great deal of English Game ancestry, along with Faverolles and the heavy breed Coucou de Malines.

The Marans made its official debut in 1914, at the national exhibition in La Rochelle, France, under the name of 'Country Hen'. However, it was due to the hard work of a Madame Rousseau during the early 1920s that the breed started to take on a standardised form, complete with the now renowned dark brown egg colour.

In 1929, some Black, Cuckoo and White Marans were imported into England by Lord Greenway. The type he selected to breed were clean legged, and this was the type that became standardised in Britain. Today, the usual colours for the British strains are the Dark and Silver Cuckoo, with Whites and Blacks being far harder to locate.

The Bantam Marans was developed during the 1950s and should also carry the dark egg trait.

Appearance and temperament

Marans are a good, robust, heavyweight breed that should cause few problems. The birds are docile and not prone to doing anything exciting or out of the ordinary, being content to be relaxed garden hens. They would make a very good choice for the novice keeper. The chicks are fairly easy to sex at an early age, as the males tend to be lighter in colour and carry a larger white head spot than the females. Spare males will grow on into useful table birds, but this can take rather a long time. Bear in mind that

This hen demonstrates the colour called cuckoo.

Left: *This Marans British Large Fowl is more of a utility shape. The show birds tend to be a bit heavier.*

they are great consumers of large quantities of food before they reach a useful table weight.

Acquiring a strain that lays eggs with the darkest brown pigment is usually achieved at the expense of egg numbers. You may wish to make a small compromise if you want a quantity of eggs, but never settle for a bird that lays pale eggs. When seeking a new cockerel to breed from, it is best to buy a clutch of really dark brown eggs from a breeder. One of the chicks from these eggs is bound to be a cockerel and by acquiring your new cockerel this way, you know he is from a dark egg strain.

In France, the Marans breed still carries the feathered legs of its Langshan ancestors, and became standardised in 1930. In this form, the most popular colours are the Copper Black and the Wheaten.

The Copper Black takes its name from the orange neck hackle.

The French version of the Marans has been in the UK for only a few years, but is already well regarded and very popular, often producing eggs of an even darker shade than those found in the British strains.

Points to bear in mind

Do not confuse the commercial breed called a 'Speckledy' with the true Marans; buy your stock from a reputable breeder.

In the French strain keep an eye out for scaly leg, but otherwise this breed should pose no problems to the novice.

Above: *A British Marans Bantam hen. A Marans hen has to lay a dark brown egg, as without that trait it really cannot call itself a Marans.*

Left: *A French Copper Black Marans hen, showing the feathered legs found in this variety.*

New Hampshire Red

This poultry breed began its development around 1915 and was created entirely from strains of the already popular Rhode Island Red. It took a while for a consistent breed type to develop, with the breed being admitted to the American Standards of Perfection in 1935.

New Hampshire Reds were imported into the UK as early as the 1940s, but have only recently been gaining in popularity with poultry keepers who appreciate their practical qualities. A very active breed club exists, so although it is only found in small numbers, the breed does not fall under the remit of the Rare Poultry Society.

Appearance and temperament

The New Hampshire Red was designed to improve certain existing traits in the Rhode Island Red, such as early maturing and fast feathering. It was also designed specifically to improve the bird's table qualities and produce larger eggs. Over years of intense breeding, the colour changed from the deep, dark red found in the breed's forbears to a much lighter and brighter shade. It

is desirable for the males to have a line of reddish pigment down the sides of the shanks.

New Hampshire Reds are an exceedingly practical breed for beginners, as they lay well, are genetically strong, with robust chicks, generally healthy and neither flighty nor lazy.

New Hampshire Red Bantams were created during the 1940s, but it is reported that instead of developing them solely by selecting from the Rhode Island Red, they also included Brahma and Wyandotte in their makeup. They are a useful utility breed of Bantam, with good egg-laying ability. The hens make only mediocre broodies and mothers.

Below: *New Hampshire Red Bantams are fast proving themselves as a practical utility breed of bantam suitable for back gardens.*

Norfolk Grey

The Norfolk Grey was designed as a dual-purpose breed, with a shape that was supposed to look like a cross between a Wyandotte and a Dorking. It was created by Fred Myhill and first exhibited in 1920. Its original name – the Black Maria – was an unfortunate choice, as it was also the nickname given to a type of police vehicle used for transporting prisoners. By the middle 1920s the name was quietly changed to Norfolk Grey.

The true breeds used in the birds' makeup will probably remain unknown, but Grey Old English Game, Leghorn and Black Orpington are likely to have figured strongly.

The Norfolk Grey has never been a popular breed, except amongst people who live in that county, and only a few years after its debut was already becoming scarce.

Appearance and temperament

There is only one colour, namely a glossy black body plumage with the hackle, saddle and back feathers a clear white (known as silver) with a black centre striping. In the females, all the body colour remains black, with only the neck hackle white with the black striping.

Although the standard calls for adult males to weigh around 3.6 kg, only the very best specimens seem able to attain that weight. For a bird classed as a Heavy Breed, many are rather too small.

The breed is a perfectly practical one. Owners report that it lays reasonably well and is an active forager. A good Norfolk Grey can easily win rare breed classes at shows, but locating good-quality stock can be difficult. Otherwise, this breed should give no problems.

Above: Norfolk Grey female. The breed makes a good practical introduction to keeping rare breeds.

Above: Unlike the very similar Birchen colour, Norfolk Greys should display no lacing on the breast.

North Holland Blue

The breed type known as the North Holland Blue in the UK could easily be classed as something of a mistake. This breed was introduced into the UK in 1934 from the Netherlands. This was very close to the time that the similar-coloured Marans was first imported and, unfortunately, some confusion seems to have occurred when the standards were created. This resulted in the Marans, which have feathered legs in France, being standardised without feathered legs in the UK; and North Holland Blues, which are clean-legged in Holland, becoming standardised with feathered legs in the UK.

Below: The legs of the North Holland Blue should only be lightly feathered. The breed is rare and in need of more support.

It might have been the stubborn British temperament that refused to accept this error, but the result was that these standards were never changed and the breed today in the UK remains one with feathered legs. It is indeed a unique form of the breed and only found in the UK in this form.

The breed was fairly popular during the Second World War, when flocks of around 500 could be found on farms throughout the South of England. The breed was mostly used as a table bird, but as with most pure breeds, its popularity waned as new commercial hybrids came onto the market.

Appearance and temperament

The UK version of North Holland Blue is a placid fowl, well marked and a layer of a reasonable number of eggs. It makes a good back-garden breed. As there are exceptionally few of them left in existence, it would help increase numbers if more people decided to keep and breed them as a practical utility bird, as this was the purpose for which they were initially created. They are suitable for the novice keeper.

The small walnut comb should resemble a raspberry cut in half. Sometimes this comb also sports little bristly hairs.

Orloff

Originally found in the Gilan area of Northern Iran, the Orloff takes its name from the Russian Count Orloff-Techesmensky. First exhibited at a poultry conference in St Petersburg in 1899, the birds were later exported to Europe and the USA. However, the writer Houdekow in 1774 claimed Orloff were already well known in Russia under the name of Chilianskaia. Other Russian breeds of fowl recorded at about this time sported crests and beards and went by the name Siberian fowl, but these were not the same as the Orloff.

Appearance and temperament

The Orloff is a striking-looking bird. The standard calls for a gloomy and vindictive expression, the result of the large, overhanging eyebrows, curved beak and large beard and muffling. However, the birds themselves are very far from gloomy. In fact, they are rather bouncy characters, exhibiting a great deal of curiosity and intelligence and often the first to investigate something new in their environment. The early hatched pullets that come into lay in October should continue to lay throughout winter.

The most frequently found colour is the Spangled, although White, Mahogany and occasionally Cuckoo and Black have been exhibited.

The Orloff is classed as a rare breed, but the numbers of the Large Fowl are currently increasing, although Bantams are still very scarce.

Left: The Orloff should have an upright carriage, and look powerfully built. The neck should be heavily covered in feathers (called a boule). This makes the neck look very thick.

Points to bear in mind

The Orloff should cause few problems for the novice. When carrying out routine handling, remember to check the beard and muffling for mites. As Orloffs exhibit a strong flock tendency, exercise extra vigilance to prevent bullying when introducing new birds.

Orpington

If the 1850s could be described as the years of 'Hen Fever' associated with the Brahma and Cochin breeds, then maybe the early years of the twentieth century can be called the hen fever years of the Orpingtons. It was reported that at a New York Show in the early twentieth century the sum of $2,500 was paid for the winning Orpington Hen. Luckily, we no longer have to pay such exorbitant prices as realised at that show, but the Orpington breed is still hugely popular today, with stock from well-known breeders always very much in demand.

Right: Buff Orpington hen. The Buff variety carries less fluff around the thigh area than the other colours.

The Orpington as a breed has a very well-documented history. The first to be created was the Black, launched by William Cook of Orpington, Kent in 1886. The Langshan, Plymouth Rock and Minorca were used in its makeup. Cook himself extolled the virtues of the Langshan-Minorca crosses in his books. The Black was closely followed in 1889 by the White, and the well-known Buff arrived on the scene in 1894.

Within four years of its launch, the Black was being changed in shape by another breeder called Joseph Partington. These changes, which added a wealth of feathering, were something Cook was none too pleased about, as shown in his book *Poultry Breeder and Feeder* "If judges would but study the laying qualities and general shape of the fowls, and not spend all their time on judging the plumage, it would be more beneficial to poultry keepers at large." Yet the public loved the fluffy plumage and that is the standard that stayed. However, the original Black Orpingtons were not lost; they were

Left: Buff Orpington male. The Orpington is a huge, placid breed.

Above: Lavender Orpington female. The Lavender colour is not yet standardised, but it is already becoming very popular.

even heavier. A Bantam form is available for those who feel their space is too limited to accommodate the Large Fowl.

The number of colours has increased dramatically, although not all of these are yet standardised. The most recent creations are the Gold-Laced, the Silver-Laced, the Lavender and the Chocolate. The chocolate gene was first discovered by the late Clive Carefoot, a well-known poultry geneticist and author. In 1995 he wrote "I hope the gene will be preserved, because it would look nice on the Orpington shape" and indeed it does, but at present it is only found in Bantams.

The Lavender colour has been a deliberate creation, refined over a number of years by Priscilla Middleton of Tonbridge, Kent, and Spangled Lavenders are now starting to make an appearance.

Points to bear in mind

When breeding Orpingtons, you may need to trim away the feathers around the vent area as an aid to fertility. Houses will need large doors and nest boxes and you should also provide low perches. Owners may suffer aching arms from trying to carry these gentle giants any distance, but otherwise the breed is an ideal introduction to the joys of poultry keeping.

refined as a practical breed in Australia and then returned to Britain as the Australorp (see page 123).

By 1891, Black Orpingtons were being exhibited in the USA and 'Crystal White' Orpingtons became well known as a utility breed there. The Buff Orpington did not have the same history as the Blacks or Whites, with Cochin, Hamburg and Dorking being claimed for their origins. However, many people, including the illustrator and writer Harrison Weir, strongly believed that they were mostly created from a good farmyard breed known as the Lincolnshire Buff (see page 147). Other colours of Orpington created around the same time were the Cuckoo, Jubilee, Spangled and Blue.

Appearance and temperament

The Orpington found today is a very large placid breed, with a wealth of feathering. When buying Orpingtons, look out for a really big bird, with adults weighing around 4.5 kg. The Blue variety should be

Right: Spangled Orpington female. In other breeds this black-and-white colour is called Mottled.

Plymouth Rock

In 1847, Dr. John C. Bennett gave the name 'Plymouth Rock' to a breed of poultry he had been developing in the late 1840s. Glowing reports of the breed appeared in the newspapers of the time (mostly supplied by himself or his close friend George P. Burnham). However, reading between the lines it becomes clear that the so-called 'breed' was not as uniform in type as was initially claimed. Much mention was made of the fact that it had either five toes or four, and blue, green, white or yellow legs, with the shanks being either feathered or clean. Daniel Jay Browne in the *American Poultry Yard* was rather forthright, referring to the breed with the words: "Plymouth Rock - This is the name of a mongrel breed of some notoriety, lately produced by Dr. J. C. Bennett, of Plymouth, Massachusetts."

Dr. Bennett's strain of Plymouth Rock soon died out, as the craze for Brahmas was just starting at that time, and both Burnham and Dr. Bennett were very much engaged in producing large numbers of Brahmas for the demanding market created in the Hen Fever years of the 1850s.

Although the 'breed' was no more, the name persisted and in 1869 D. A. Upham exhibited a trio of barred-coloured fowl given the name 'Improved Plymouth Rocks'. These birds made such an impression that Upham immediately received orders for over 100 sittings of eggs at $2 a sitting. Unlike the earlier Bennett strain, Upham soon had his strain of Rocks breeding true to type.

Many other breeders were also working on very similar-coloured fowl at the time, and all these eventually came together under the name Barred Plymouth Rock. It is generally accepted that all owed their colour to the Dominique (see page 132-133). After the Barred variety, came the Buffs, Whites, Blacks, Partridge, Blue and many other colours. Breeders of 'Rocks' as they are often simply referred to, tend to be rather fanatical about the intricacies of the colour of their birds. Many long and often heated discussions take place as to the exact shade of colour that a Buff Rock should be and the width and preciseness of the colouring of the barred plumage pattern.

Right: *Buff Plymouth Rock Bantam. The shade of buff is important; it has to be an even gold-buff right down to the skin.*

to look after, as well as robust and hardy. The Large Fowl has been quite scarce for a number of years and unfortunately egg production can be rather low in some strains.

Points to bear in mind

This breed should cause the novice no practical problems. The barred variety is often doubled-mated to achieve the desired colour on both cocks and hens. Do not confuse the well-known commercial hybrid called a 'Black Rock' with the pure black 'Plymouth Rocks'.

Above: White Plymouth Rock Large male. The legs should be bright yellow. To keep the legs the right colour, Plymouth Rocks need access to plenty of grass.

For the fancier, breeding Rocks is an exacting science. For the backyard poultry keeper this perfectionism means that slightly mis-marked but otherwise excellent quality stock is available from breeders. These birds make a useful and pretty introduction for novices who are not interested in breeding for show.

The Bantam form is now far more popular than the Large variety, being reasonable layers and easy

Left: White Plymouth Rock Large female. The Large Fowl are not as popular as the Bantams.

Rhode Island Red

The Rhode Island Red was a deliberately created breed, developed on the poultry farms in the Little Compton district of Rhode Island, USA. Its development began around the mid 1850s. Although the breeds used in the first experimental crosses were Malay and reddish-coloured Shanghais, it was said that at one time every great breed of poultry ever known in America found its way into the flocks of the Rhode Island farmers. Brown Leghorn and Indian Game were recorded as being used in later breeding strains.

It was only from around 1895, when a poultry breeder named Dr. Aldrich started to refine the breed by careful selection of quality, red-coloured fowl found in the Little Compton area, that the breed really took shape. It was shown under the name of Rhode Island Red at the Providence (Rhode Island) show in 1895 by Richard V. Browning of Natick, Massachusetts. However, the breed only became truly standardised in 1904. During the early part of the twentieth century, it rapidly expanded to become one of the best-known fowl in the world.

The breed possessed so many good qualities that it would have been difficult for it to have remained in obscurity. In the days before treatments such as antibiotics, a breed with proven vigour, robustness, strength and ease of reproduction, combined with good utility merits and good feed conversion abilities (not to mention some good publicity) was bound to be a winner.

If you ask anyone to name a breed of chicken, the chances are that fairly quickly someone will mention Rhode Island Red. Unfortunately, this has also proved a problem for the breed, as most people automatically assume that any breed with red plumage must be a Rhode Island Red.

Appearance and temperament

A few points set a good Rhode Island Red apart from other, more basic red-coloured breeds and some commercial hybrids.

Firstly, the breed should be a deep, rich, red colour – not the pale reddish buff colour found in most commercial hybrids. If you part the feathers, the colour of the underfluff should also be red or salmon-coloured – not whitish or smoky. A Rhode Island Red should be a large bird. A hen should weigh around 2.5kg and the male no less than 3.6kg.

Right: *The colour of the underfluff in a Rhode Island Red should be red or salmon-coloured; it should never be pale or white. The legs should be yellow. This is a Bantam male.*

Right: *This Bantam female shows the deep broad shape required.*

Rhode Island White

The Rhode Island White is little heard of outside its native America. The breed originated using Rose-combed White Leghorns, White Wyandottes and Partridge Cochins, and was the work of J. Alonzo Jocoy from Peacedale, Rhode Island. Between 1918 and 1920, the Rhode Island Whites won a number of prestigious egg-laying trials, proving their merit. Subsequently, this breed provided the basis for a number of commercial hybrid egg-laying strains.

The shape of the body should be deep, broad and long, giving the birds an oblong appearance; they are often referred to as brick-shaped. A Rhode Island Red should have yellow legs. Although some allowance can be made for in lay hens and adult birds, this colour should still be visible.

Show strains of the Rhode Island Red are frequently a very dark Chocolate colour, but strains of Rhode Island Red that have been kept pure for utility qualities are lighter in colour. However, they should still be distinguishable from the commercial hybrid poultry.

Although mostly found in the single-combed type, the Rhode Island Red is occasionally found in a rose-combed variety. This very early type was admitted to the standards in April 1905. Bantam versions of both types were developed during the 1920s.

The Rhode Island Red has often been used as a cross with the Light Sussex. Chicks from this cross produce good utility birds that can be sexed at hatching, due to the gold gene carried by the Rhode Island Red and the silver gene of the Light Sussex.

The Rhode Island Red was used to create another very useful sex-linked breed called the Rhodebar. This was developed in England in 1947 by crossing imported Danish stock with the Golden Brussbar. It retained all the Rhode Island Red's utility qualities, with the advantage that breeders could tell the sex of the chicks at hatching.

Below: *Show type strains of Rhode Island Red are an exceptionally dark red colour.*

Sulmtaler

The Sulmtaler hails from an area near Graz in Austria. It was developed from local fowl crossed out to Houdan, Cochin and Dorking during the late 1860s, and then crossed back once more to the local breeds found in the traditional poultry rearing area of Stiermarken. The breed was then standardised into a predominantly table fowl, with useful egg-laying qualities.

The Sulmtaler first appeared at UK shows in 1991 in Bantam form, and Bantams continue to dominate the show scene today. Large Fowl are very scarce in the UK.

Above: The body of the Sulmtaler should be broad and cobby in appearance. Large Fowl, such as this male, are far scarcer than the Bantams.

Appearance and temperament

The term 'broad' appears frequently in the standard for this strong, cobby-shaped breed and it is worth remembering that not only should the back be broad, but the birds should have a broad, deep and well-rounded breast and abdomen.

The Sulmtaler is found in one colour only – the Wheaten – where the males show the traditional Black-Red colouring and the females a soft Wheat body colour.

The hens have an unusual S-shaped comb (called a wickel), where the front and back parts fall to opposite sides of the head. Both males and females display a small crest behind the comb.

It is unfortunate that many experienced people overlook this breed, thinking it resembles a cross-bred Silkie, with its small crest and cobby shape. However, the Sulmtaler has

many virtues. The Bantams are very good egg layers, the breed is docile and easy to keep and very hardy. Fertility is usually good, with strong robust chicks. The hens make good broodies and mothers.

Points to bear in mind

Locating Large Fowl stock may be difficult, although the Bantams are more readily available. Otherwise, this breed should pose no problems.

Left: The female's combs fall to both sides of the head in an S-shape, when viewed from above.

Sussex

Throughout the nineteenth century, the area known as the Weald of Sussex was renowned for sending good-sized, plump birds to the tables of London on a regular basis. Although many people regarded these 'Kent' or 'Old Sussex' fowls as just a type of crossbred Dorking, the truth was that over the years the farmers of Sussex had been quietly developing a separate breed that could be produced to a very consistent standard.

The story of how the Sussex breed came to be standardised is well known but worth repeating. In early 1903, the author Edward Brown addressed a meeting of some Sussex farmers at Lewes and bemoaned the fact that they had permitted the fowl, which had been previously native to Sussex, to die out. One of the farmers present at the meeting, Mr E. J. Wadman, promptly replied that he had been maintaining a flock of pure Red Sussex for several decades and still had them. Within the year, a Sussex Poultry Club was formed and by their diligence the members firmly put the Sussex fowl back onto the map of British poultry breeds.

Above: When looking at a Light Sussex, note that the black centre of each hackle feather should have a clear white margin all around it.

Appearance

The first colours to be accepted into

Right: Speckled Sussex Large Fowl female. This was one of the first recognised colours in this popular breed.

the new club were the Light, the Speckled and the 'Red or Brown'. In 1913 the Red and the Brown colours were separated, with a precise standard for each. The Red, Brown and Speckled have gone through some rather precarious patches over the years. The Red and Brown in particular are now seldom seen except at the larger shows and are sorely in need of more support, being found in fewer numbers than are many rare breeds.

The Speckled Sussex is very attractive, with a deep mahogany body colour and feathers tipped with a bar of black and a spot of white, the black part having a lustrous green sheen to it.

The Light Sussex, with the Columbian pattern of black and white, is by far the best-known variety and has never faded in popularity, partly due to the fact that it carries a silver gene. This means that it can be used with Rhode Island Reds to produce quality utility birds that can be sexed on hatching because of the different down colour in male and female chicks. For many years the Rhode Island Red x Light Sussex was the mainstay of the British poultry industry, providing males for the table and females for egg production.

The Coronation Sussex is similar to the Light, but a rich pigeon-blue colour replaces the black.

The Buff Sussex was developed during the 1920s and has made something of a comeback in recent years. The combination of a rich beetle-green sheen on a buff background is very pretty.

The Silver Sussex was developed in 1948 by Captain Duckworth. The body colour is

Left: The Buff Sussex is basically the same Columbian pattern found in the Light variety, but with a bright buff background instead of white. This is a Large Fowl female.

The Sussex is only found in a single combed variety.

Right: Red Sussex Large male. Although one of the first colours to be standardised, the Red Sussex is now very scarce.

mostly black, but with a white hackle striped with black. The wing bows and back are both silvery white and each feather of the black breast has a white shaft and is laced with silver.

Over the past century, the Sussex bred for exhibitions took on a heavier appearance, with size, shape and colour being strongly selected as a point of merit. Concentration on show points did lead away from the practical aspect of the breed, and although show birds look exceedingly attractive, they are no longer classed as good laying fowl.

A few strains of both White and Light Sussex have been preserved solely for their utility merits and are once again becoming more popular with people who are looking for a practical backyard bird, with the bonus of a well-documented UK history. It is ideal for the novice.

A Bantam Sussex was developed around 1920 and is just as popular as the Large Fowl, being a good utility type of Bantam.

Transylvanian Naked Neck

Many people assume that the Transylvanian Naked Neck acquired its name due to the vampire-attracting features of a bare neck. However, it appears that the breed did actually originate from that area of Europe. Although folk tales abound telling how the breed lost its feathers scavenging in compost heaps or because of accidental scalding, the truth is that the absence of neck feathering is due to a specific gene also found in other breeds, including the Malgache Gamefowl of Madagascar and the Ga Don of Vietnam.

Appearance and temperament

Only a devotee could ever describe this breed as beautiful. The Victorian writer Lewis Wright was obviously not a fan; in his *Illustrated Book of Poultry* he describes their appearance as being "peculiar, but most unpleasant." Naked Necks are not just missing feathers on the neck. If you lift the body feathers you will find that there are featherless areas on the body as well.

The breed has many merits. It is a very good layer of large eggs and an excellent broody. It makes an exceedingly good meat bird; indeed, the breed is used extensively on the Continent for this purpose. It tolerates changes of temperature well. In short, the Naked Neck is a very hardy and thrifty fowl, but it can take a very keen poultry keeper to overlook the appearance in favour of its practical merits. It is found in both Large and Bantam forms.

Points to bear in mind

The breed should cause no practical problems for the novice. However, persuading your friends and neighbours that the birds really should look as they do, and that you have not been plucking them, might prove more difficult.

Right: The breed is popular in the USA where it is called the Turken.

Left: Despite its lack of feathers the breed tolerates low temperatures well.

Wyandotte

The Wyandotte is an American breed and a very large family indeed, with numerous colours now standardised and a readily available Bantam form. From its conception, it was always intended to be a pretty breed. American breeders had seen the diminutive Sebright with its fancy lacing and set about creating a large, practical fowl that would look just as attractive. Indeed, in the early years, the breed was known as the American Sebright. The Silver-Laced Wyandotte was slowly developed between 1864 and 1872, but only became sufficiently uniform in type to be admitted into the American Standards of Perfection in 1883 under the name of Wyandotte, a name that the Huron Native Americans called themselves.

The name Wyandotte was not greeted with much enthusiasm in the poultry world. As C. J. Ward, the editor of the *American Poultry Journal* wrote at the time: "American Sebrights were christened Wyandottes, which we think absurd and nonsense, as it means nothing and will cause confusion, but it is done, and so we will all say 'let it go.' "

A Gold-Laced variety followed soon after and so did other laced varieties, such as Blue-Laced and Buff-Laced. Then it was the turn of the Pencilled varieties to come to the fore. The delicate pencilling found on the Cochin and Dark Brahmas was a very popular colour at that time and it made sense to place this pretty colour on a rather more practically shaped breed of fowl.

Appearance and temperament

The Wyandotte is a very hardy, robust, no-nonsense sort of chicken. It is certainly one of the most popular breeds in the UK today, finding favour with show exhibitors and backyard poultry keepers alike. With its attractive markings, compact, neat head points and tidy body shape, it is often the breed that encourages the novice, previously content with a commercial hybrid, into the world of fancy poultry. There are certainly enough colour varieties to keep the most ardent Wyandotte fancier happy for many years, with 14 colours currently standardised in the UK.

The breed is available in both Large Fowl and Bantam, and it is often the Bantam that attracts initial attention. Being a popular breed, the Bantam

Below: *Lavender pair. Being a newly created colour, the Lavender still needs some work before it is truly perfected. The tails are still rather large.*

Right: The flecking on this Partridge Bantam male's breast shows he was produced from a pullet breeding strain. Males from a cockerel breeding strain have a solid black breast.

important, with the leader (the back spike of the rose comb) following the line of the head, it should never be a 'flyaway' comb that sticks out into the air.

The Wyandotte should have yellow legs, although a small allowance is made for fully adult birds and those in full lay, where the pigment may fade a bit. To get good yellow legs in the Black variety can prove exceedingly difficult, because the gene for the yellow leg is somewhat incompatible with the best solid black coloration. Breeders often resort to double-mating to obtain the correct colour on the legs, using pullets with perfect leg colour and males with a slightly lighter undercolour.

Wyandotte is usually found in reasonable numbers at most poultry shows throughout the country. The Wyandotte is a very curvy breed; a fancier can instantly recognise it by its rounded outline alone. The females make very good broodies and mothers and are often used for hatching eggs from other breeds.

Showing

For show purposes, the colour pattern is very precise. Show exhibitors will almost certainly practice double mating (see page 76) for both the Laced, Partridge and Pencilled varieties. A novice wishing to take up one of these colours for breeding or showing would be well advised to seek out a 'pullet breeding' strain, as it is far easier to sell on spare stock produced from these than it is from the 'cockerel breeding' strains. When buying your initial stock, do be sure to ask an experienced Wyandotte breeder to point out the differences in coloration that will be required in a pullet-breeding male bird.

The shape of the comb in the Wyandotte is very

Points to bear in mind

This breed is an excellent choice for the novice and should cause no practical problems. Sourcing certain colours in Large Fowl may prove difficult, as a number of the once popular colours are becoming quite rare in this size.

Right: Wyandotte Black pullet. The comb of the Wyandotte always fits closely to the head.

Left: A young Barred Bantam pullet, showing the neat, clear barring found in the breed.

Introduction to True Bantams

In William Pulley's *The Etymological Compendium* of 1830, he states "The small fowl, designated by the name of Bantam, derives its appellation from Bantam, in the Isle of Java; and was first introduced into this country in 1683, when an embassy arrived in England from thence."

For many years, if anyone asked for a definition of a Bantam, the standard reply was "A small fowl from Java". However, things have moved on in terms of clarity and we now distinguish True Bantams from the miniatures of the large breeds.

Unlike the miniatures of large breeds, which should be around a quarter the size of the Large Fowl equivalent, a True Bantam has no Large Fowl counterpart; it only exists in the miniature form in which it was first created.

The True Bantams are a very select group, developed with no other purpose in mind than to be ornamental creatures. They were required to be neither a laying fowl nor a meat provider. In short, the True

The Sebright has an elaborate plumage pattern.

The Dutch is one of the most vigorous of the True Bantams.

Bantams have no real purpose other than to be elegant and charming – and this is something they really do excel at.

True Bantams can require more care than the large breeds, and many have certain inherent traits that need special attention. However, these problems are more than offset by the fun of owning these little gems. From the delicate markings of the strutting Sebright to the rounded fluffy body of the good-natured Pekin, these little breeds are only in existence because we humans find them charming little individuals. And there really could be no better reason than that.

Barbu d'Anvers

By 1888 the Barbu d'Anvers was already of a type not far removed from the one we have today. In 1911 a consignment of Belgian Bantams was displayed at the Crystal Palace Show. The birds attracted much interest and it was from that date that the UK history of this breed starts. It is also known as the Bearded Antwerp Bantam (Barbu meaning 'bearded').

Appearance and temperament

This breed should always be as small as possible. Most of what looks like a thick neck is actually made up of feathers in a 'boule'. The neck hackle of the male should resemble a cape. The birds have a rose comb and need to sport a wealth of beard and muffling. The face should have an 'owl-like' appearance.

The colour most widely encountered is Quail, a combination of golden buff and black, although many other colours are available.

This little bird is a true dandy of the poultry world. The males think they are the biggest, most elegant and important birds in the entire poultry yard; they will proudly strut and posture to prove the point. This has made them superlative show birds, as they are real egotists. The females are very sweet natured and tame easily.

If Barbu d'Anvers are kept for showing, they are best housed in small runs and pens. However, despite their small size, these birds are not delicate and do well on extensive range, provided they are protected from predators.

Above: *Barbu d'Anvers Blue Quail male. The leader of the rose comb should follow the line of the neck. The wattles should preferably be absent.*

Points to bear in mind

The cockerels have a well-founded reputation for aggression towards their human keepers, so it would be best to keep the males away from very small children until you have established the temperament of your bird.

Chicks are very small and need careful attention. This breed is prone to suffering from Marek's Disease, which causes paralysis. Some breeders vaccinate against Mareks Disease, so do speak to the breeder about this before buying, as certain strains are more prone to the disease than others.

Left: *Barbu d'Anvers Lavender Quail male. In the breeding season the males will be protective of their hens, but females are always sweet natured.*

Barbu du Grubbe and Barbu de Watermael

The Barbu de Watermael is a variety of the Belgian d'Anvers. It sports a crest, as well as the usual beard and muffling. The Barbu de Watermael is often exceedingly tiny and lightweight. It is only found in limited numbers in the UK, but its diminutive size means it can be kept in the smallest of gardens, with pairs often living happily in large rabbit hutches, with the occasional outing into the garden. Because the chicks of all the d'Anvers breeds are very small, it is always wisest not to risk rearing them alongside chicks of Large Fowl varieties as they easily become trampled.

The Barbu du Grubbe is a rumpless variety. Although otherwise identical to the d'Anvers, it lacks the vertebrae of the tail and hence the tail feathers are missing. This breed may require a little more attention, as rumpless birds can often suffer from fertility problems.

The crest should not cover the eyes.

Left: *The Barbu du Grubbe is one of the few breeds of poultry that is lacking a caudal appendage, or 'parsons nose'.*

Above: *The Barbu du Watermael is not as frequently seen at shows as its cousin the Barbu d'Anvers.*

This Barbu de Watermael chick is already showing signs of the crest and beard.

Barbu d'Uccle and Barbu d'Everberg

This very small breed originates from the municipality of Uccle, at the southeast border of the Brussels-Capital Region in Belgium. It is believed to have been developed in the 1890s from the Belgian d'Anvers and the Booted Bantams, which were already in existence at the time.

Appearance and temperament

Barbu d'Uccle should have a wealth of muffling and beard, small or preferably absent wattles, a single comb and feathered legs and feet.

In the Millefleur (meaning a thousand flowers), the ground colour is a rich mahogany, with the feathers ending in black, followed by a white tip. The correct pattern is very hard to get right for show. A dilute form of this colour is called Porcelain. In these birds, the background is a pale straw colour and the black is replaced with a soft lavender.

If you plan to keep this breed for showing, it is worth noting that it improves as it gets older. This is especially true of birds with the attractive Millefleur and mottled colours. It is not uncommon to find a three- or four-year-old bird winning the prizes.

The Barbu d'Uccle is hugely popular. The females are amazingly sweet tempered and can become exceptionally tame. D'Uccle males can prove

Right: The Barbu d'Everberg is a rumpless version of the d'Uccle. It is seldom encountered, but kept by a few devoted followers.

aggressive towards humans, so it would be best to keep the males away from very small children until you find out the individual's temperament.

If you are not keeping the birds for showing they will generally do well on extensive range, despite their feathery legs. Do be sure to protect them from predators, as small bantams can fall easy prey to hawks.

Points to bear in mind

Unfortunately, this breed can be prone to Marek's Disease, which causes a form of paralysis. Some breeders vaccinate against it and it is wisest to seek advice from a breeder before you buy your birds, as the disease does seem to affect certain strains more than others.

When buying, check for mites in the beard and muffling. Any signs of scaly leg will need treating.

Provide extra perch space to accommodate the birds' feathery feet, and do keep these chickens off muddy ground.

Right: The white areas of the Millefleur plumage will increase as this Barbu d'Uccle hen grows older.

Booted Bantams

Bantams with feathered feet have been well known since the very early years of poultry keeping, being mentioned as far back as 1600. In 1857, Edmund Dixon wrote "The feather-legged bantams are now as completely out of vogue as they formally ‹sic› were in esteem, we ought perhaps to have referred them to the anomalous fowls." Thankfully, this is something that cannot be said today.

In the UK, Booted Bantams are classed as a rare breed and a True Bantam. However, they are a popular breed on the Continent, especially in the Netherlands and Germany. The breed is also called 'Sabelpoot', meaning 'sword leg', a reference to the stiff feathers on the hocks.

Appearance and temperament

Many people confuse the Booted Bantams with the Belgian Barbu d'Uccle, but take a look at the face and you will see that unlike the d'Uccle, the Booted Bantams do not have beards or muffling.

Booted Bantams are sprightly little birds. Keeping the foot feathering clean can pose a problem, but they take to confinement well. However, if you are not keeping them tidy for a show, the birds do

Below: Blue-booted Bantam Lemon Millefleur. There should be plenty of foot feathering,

There should be feathers on the middle toe.

well if allowed freedom. This breed is found in an interesting and pretty range of colours, such as Buff Mottled, Lavender Mottled and Silver Millefleur.

Points to bear in mind

Unfortunately, Booted Bantams are prone to suffer from a Marek's Disease, which causes paralysis. Obtaining your stock from a breeder who vaccinates against this disease is a possible option, but discuss this with the breeder, as certain strains are more susceptible than others. They can also be prone to scaly leg. Provide extra perch space to cater for the foot feathering.

Although all these things can prove problematic, the very placid nature of this breed makes it suitable for the attentive novice.

Right: Unlike the Barbu d'Uccle, the Booted Bantam has no beard or muffling.

Dutch

This is one of the smallest breeds of poultry in the UK. Males weigh around 550gm maximum and the females are even smaller. They were standardised in the Netherlands in 1906, but only arrived in the UK in the late 1960s, when they went by the name 'Old Dutch', later shortened to Dutch.

Dutch take to confinement easily and are often kept in small enclosures. Pairs are often even kept in large rabbit hutches. This has made them a very popular back garden breed, especially in the Netherlands.

This ornamental True Bantam breed is a very robust little bird that tends to suffer from few health complaints and makes a very good introduction to the True Bantams.

Appearance and temperament

The breed comes in such a huge assortment of colours that if a breeder became bored with one colour they could quickly move on to another. Devotees seem determined to attempt to produce it in every colour ever known to exist in poultry!

The breed description calls for the male to have an upright and jaunty appearance, with an abundant hackle and a full and well-spread tail, with well-curved sickle feathers. Dutch should have neat, white, oval-shaped earlobes that do not show signs of red in them.

Dutch will steady down to be very tame, but are not as naturally tame as breeds such as the Belgian Barbu d'Anvers or d'Uccle. They lay a surprising

Left: A Dutch pair. This breed is available in a wonderful range of colours, including this Millefleur, which means 'a thousand flowers'.

The pattern will become more pronounced with age.

The Dutch have a low wing carriage, as in this male.

number of eggs for a breed really only kept for ornamental purposes. The Dutch is a very popular show breed and appears in large numbers at all the major poultry shows.

Points to bear in mind

At first, novice breeders may be concerned about the smallness of the chicks, but will soon get used to handling them. This breed should present few difficulties, but cats and hawks can be a problem if the breed is left to run loose.

Right: Lavender Dutch male. Lavender is a true breeding form of the blue colour. In Europe it is called Pearl Grey.

Above: Dutch Gold Partridge pullet. This is the colour most frequently found in the breed.

The earlobes should be white with no trace of red.

Below: Dutch Blue Yellow Partridge pullet. In this colour a subtle shade of blue replaces the black colouring.

172

Japanese

This small breed is also known by the name of Chabo. It originated in Southeast Asia and was depicted in Japanese artwork from the early 1600s. In his picture *The Poultry Yard,* painted in 1660, Jan Steens depicts a small buff bantam cockerel with very short legs and an upright tail. This bird strongly resembles the breed we know today.

On first seeing the Japanese breed, the casual observer can be forgiven for thinking that it appears to be sitting down; only when it chooses to walk off does the truth dawn; yes, its legs really are that short! When buying, look for a bird with very short legs and an upright tail. The main sickle feathers on the male should stand straight up like swords.

The Edwardian writer Edward Brown described the birds as "grotesque in the extreme", but as we know, beauty is in the eye of the beholder and this breed has many devoted followers throughout the world.

Appearance and temperament

The Japanese breed is found in a wide range of colours, and Frizzle and Silkie-feathered varieties are also available. In Japan, the breed is found in various forms, some of which have far larger combs than are seen in the West. The Japanese is a very well-behaved breed and the birds make good pets, taming easily.

Japanese can end up becoming very dirty and soiled unless kept indoors. However, on fine days they love to get out and enjoy being a typical chicken, even if they do have to look at the world from a rather low vantage point.

Points to bear in mind

Because of the short legs, fertility can be poor and up to around 25% of the eggs will fail to hatch.

Below: Japanese Bantam female. A suitable breed for the novice prepared to provide a bit of extra care. This hen is actually standing up.

Nankin

The Nankin is one of the earliest known True Bantam breeds. It was named for its colour, which closely resembled that of the hardwearing yellow cotton cloth known as Nankeen, used in the production of breeches and waistcoats in England during the latter years of the eighteenth century.

In the 1867 edition of *The Poultry Book*, William Bernhard Tegetmeier writes "One of the most common of the old Bantams was that known as the nankin or yellow breed. These, although now seldom exhibited, still have their admirers. Their prevailing colour is that of the pale orange yellow of the nankin cotton".

The breed fell out of favour and became very scarce as early as 1902 and later was thought to be extinct. However, Mr. Martin of Wisbech, Cambridgeshire had been quietly keeping a flock for many years and these came to light in 1955. The breed is still rare and needs more supporters to keep it on a sound footing.

Appearance and temperament

This breed is permitted to have one of two comb types. The single-combed variety is most frequently seen, although the rose-combed variety does make a very occasional appearance at the larger shows.

The breed should appear neat and bold, with the breast

Below: A good example of a Nankin male. This bird displays the rich yellow colouring that gave the breed its name.

Below: The legs of the Nankin should be either blue or bluish-white.

carried well forward, a sloping back and the wings carried very low, nearly touching the ground. The legs should be either blue or bluish-white. There should be no trace of white showing in the earlobes or face.

The birds are reasonable layers for a True Bantam, and the hens make good broodies and mothers, but are not able to cover many eggs due to their small size.

Points to bear in mind

This breed should pose no problems to the novice, but take care during the breeding season as some individual males can behave a bit aggressively towards their human keepers.

Below: The male bird shown on these pages is the more common single-combed variety.

Ohiki

The Ohiki is one of the Japanese long-tailed breeds, but in addition to its long tail it also has very short legs. It was developed by mixing the popular Chabo breed, known in the West as the Japanese (see page 173) and the Onagadori (see page 121), a breed renowned for its amazing length of tail. The result is a very small, rounded bird with a single comb, short legs and very long tail feathers.

This is really only an exhibitor's bird, as a great deal of care is needed to keep the feathers of this very short-legged bird clean. The gene for the long tail will also produce a wealth of side hangings and furnishings around the tail. Anyone wishing to breed these birds needs to be dedicated and very particular about their husbandry.

Appearance and temperament

The Ohiki has become a very popular breed in its native Japan. Black-Red, Silver and Gold Duckwing are the colours most usually found. The tails of the male birds should always be long enough to reach the ground.

The eggs hatch best when set under broodies rather than in incubators. Since Ohiki are not the most reliable mothers, it will be necessary to keep additional broody breeds. They are a very tame little breed and take well to being handled.

Points to bear in mind

Keeping the bird clean and tidy is vital. Muddy ground is a disaster for this breed, which is best kept in a covered run with very deep litter. As the birds are only small, their perches need be set no higher than normal, but do raise them if the male's tail grows to such an extent that it touches the floor. Chicks can be rather weak when first hatched and may require extra attention.

Pekin

The history of the Pekin began in the Western World in about 1860, when this little breed was taken as loot from the Emperor's palace in Peking, China by British soldiers at the time of the Opium Wars. Once it arrived in the UK the breed quickly became popular, with further birds being imported in the following years.

There is no evidence that Pekins were ever produced by miniaturising the larger Cochin breed, even though on the Continent and in the USA it goes by the name Cochin Bantam. It is far more likely that the birds had been carefully refined as Imperial pets from the small feather-legged Bantams that were already in existence at the time.

Above: *Lemon Cuckoo Pekin female. This is one of the more recent colours to become available in the breed. There is no shortage of colours to choose from.*

Appearance and temperament

A good Pekin should be very short on leg length – so much so that when first observing a good example, the casual observer frequently thinks it is sitting down! A Pekin should never appear narrow when viewed from the front; it should remind you of a round ball of feathers.

The tail is in the form of a cushion of feathers that is just a curved, rounded extension of the rest of the bird. A good Pekin will display what is called tilt, whereby the head is carried a fraction lower than the cushion.

Pekin breeders seem very keen on developing new colour varieties, and the birds are now found in a huge range of colours; and if the colour you desire is not available now, there is little doubt that you will be able to find it in a few years time.

There is a version of the Pekin in which the feathers curl outwards, all of them facing towards the head. This feather type is called a Frizzle. When breeding Frizzled Pekins, be sure to cross them back to the smooth-feathered variety to maintain good

Left: *Black Pekin male. The Pekin has a cheeky attitude and can be rather opinionated.*

feather quality. Failure to do this results in a very frazzled-looking creature, with narrow wiry feathers that do nothing to keep it warm. A correctly frizzled Pekin with wide feathers is instantly reminiscent of a cute rounded hedgehog and is always much commented on by visitors.

Pekins are not destructive if allowed to roam, as the foot-feathering prevents them doing much damage to plants and borders. However, do not allow them free range during wet weather, as they quickly become muddy and dirty.

Pekins lay a reasonable number of eggs, although like other True Bantams, they are not kept either as meat birds or as layers, but purely for their exhibition and pet qualities.

Pekins make excellent broody hens and very attentive and careful mothers. They live happily in confined environments.

Although Pekins do need rather more attention on account of their feathered feet, their excellent docile temperament makes them suitable for the caring novice. They are ideal pets for children, tolerating being carried about and fussed over; indeed this little breed really seems to enjoy human company.

Above: *Lavender Pekin female. The Lavender colour is very popular, being a true breeding form of blue.*

Left: *Lavender Pekin male. Most of all, a Pekin should be short-legged and rounded in shape – never narrow. The tail forms a cushion of feathers.*

Showing

A very careful novice who buys quality Pekin stock and keeps the foot-feathering in excellent condition will usually manage to do well at shows. Pekins must be very full feathered to win top prizes as they have been refined to a high degree of perfection. The leg and foot feathering should be clean and unbroken, something almost impossible to achieve with birds running out on grass. Avoid keeping Pekins on muddy and wet ground. For showing you will need to keep them indoors on very deep, clean shavings to keep the foot feathering in good condition, and allow extra perch space.

Right: *Blue Mottled Pekin pullet. This colour can be crossed with black mottled to obtain chicks of both colours.*

Points to bear in mind

Make sure that the toes are of reasonable length as these can sometimes be shortened due to the feathered feet. All colours should have yellow legs and feet, although dark legs are allowed in the Black, providing the soles of the feet are yellow.

As with all feather-legged breeds, keep a watch for scaly leg.

When breeding Pekins, you may need to trim away the feathers around the vent area as an aid to fertility.

Left: *Pekin Cuckoo male, displaying good shape and showing the tilt required in the breed.*

Rosecomb

The Rosecomb is a flashy little jewel in the world of poultry keeping. Kept neither for its meat nor egg laying, this breed exists for visual effect only. It has retained its position as an outstanding bird, purely produced for the exhibition table, for an exceedingly long time; small, smooth-legged Bantams were mentioned by the sixteenth century.

Appearance and temperament

The fact that the breed goes by the name Rosecomb does rather give the game away. One of its most pronounced features is indeed its rose comb. This has to be a very specific shape, with a broad, flat front, tapering to a fine, even 'leader' on the end of the comb. The 'workings' (the mass of small spikes on top of the comb) must be of an even height, so that they show a basically flat surface when viewed

Left: The comb should be an even shape, with a square shape to the front and a leader that is straight and not twisted.

from the side. The earlobes should be as round as possible and must be pure white. The birds are usually found in Black, White or Blue.

The Rosecomb is not an overly friendly breed, but neither is it scatty, and with regular handling can become tame. The males can be rather aggressive towards humans during the breeding season, so it is a good idea to keep them away from children at this time.

Points to bear in mind

To maintain a Rosecomb up to exhibition standards is an arduous task. However, to keep the birds in the garden as attractive pets is not really difficult, as long as you remember that the breed is not as hardy as larger fowl and will require extra attention during the winter and in bad weather. Novices should try keeping an easier breed first.

The chicks are very tiny and should never be reared with larger chicks as they will easily be trampled. Take special care to avoid any damage to the earlobes. Good housing is vital.

Anyone thinking of keeping this breed should first ask an experienced Rosecomb breeder to point out the very many exhibition points required of the breed and the problems that may arise.

Below: Broad feathers should be evident in the Rosecomb's tail. The wings should be held rather low and not tucked up. This is a female.

Sebright

In about 1800, a group of English Poultry fanciers set out to create their vision of the perfect fowl. It was to be small, neat, precisely coloured and very attractive. But it was going to take many of them a lifetime to achieve this goal. One of the group was Sir John Sebright, a Member of Parliament for Hertfordshire, who was already well known as a livestock breeder of cattle, dogs and pigeons. By 1815, a Sebright Club catering exclusively for this small Bantam was formed. Each year, members of the club gathered to compare birds.

The precise breeds in the makeup of the Sebright were not recorded, but Sir John claimed that a buff bird acquired in Norwich (from the description most likely a Nankin) and a male bird from Norfolk that was nearly hen-feathered (showing no long sickle feathers in the tail), were included. This was probably a pit-game cock. How the lacing arrived is mostly speculation, but it is believed that a laced bird acquired from a zoo was also involved or, very possibly, a laced Poland. The Gold colour was the first to be created, followed later by the Silver. The breed is mentioned in *An Encyclopædia of Agriculture, 1826,* where John Claudius Loudon says "The other, and more scarce variety, is even smaller; and is most elegantly formed, as well as most delicately limbed. There is a society of fanciers of this breed, who rear them for prizes, among which Sir John Sebright stands pre-eminent."

Appearance and temperament

Stylised drawings by Victorian poultry artists provided examples of what the perfect Sebright would look like and breeders worked towards these exacting standards. Today, the Sebright is about as perfect as it is possible for it to become.

The Sebright is not easy to keep. To produce a bird with the correct markings, ground colour, body and head shape is very difficult indeed.

Sebrights can often prove rather neurotic, squawking at random opportunities. Because the breed is hen-feathered, there can be fertility problems. The best males exhibit no sickle feathers at all; their

The feathers are almond shaped.

Above: *Gold Sebright female. Each feather should be evenly and clearly laced with black.*

tails look just like those of the females. The gene for the rose comb is also known to decrease fertility.

Introducing new hens to a breeding pen can be very troublesome, as existing hens, in particular, can take strong exception to the arrival of any new female and will behave very aggressively towards her. For this reason, the breed is often best run in pairs. Chicks can also prove rather quarrelsome amongst themselves. As with a number of other True Bantams, some strains may be prone to Marek's Disease, which can cause paralysis. Consult the breeder before making a purchase, as vaccination is an option that some of them employ.

Taken altogether, these considerations mean the Sebright is not a breed for the novice, but with some experience, these little laced gems are a joy to observe strutting around a garden.

Above: *The male does not have pronounced sickle feathers, a feature known as 'hen-feathering'.*

These feathers are clear. Smutting in the feathers is most undesirable.

Above: *Silver Sebright male. Although the standard calls for a mulberry-coloured face, it is seldom seen in the males.*

Serama

The Serama is a new and very tiny breed of fowl developed during the 1970s and not yet standardised in the UK. It is usually claimed to be the work of Mr. Wee Yean Een, a poultry breeder from the state of Kelantan on the Malay Peninsula.

The Ayam Kapan, a type of Bantam that very much resembles the early forms of the Japanese Bantam, was one of the breeds used in the makeup of this new breed. Bear in mind that many Malaysian breeds of poultry begin their name with the word Ayam, this being the Indonesian word for chicken. The Kapan has a high chest carriage and longer legs than the Japanese (Chabo – see page 173). Morever, it in no way, shape or form resembles the Modern Game Bantam, as is often claimed.

Some versions of the breed's history say that Ayam Kapans were out-crossed to a Silkie in the hopes of producing a small Silkie fowl, but the idea of a small Silkie was abandoned when the first chicks that hatched did not contain Silkie-feathering. The Japanese Bantam played a major part in the creation of this breed, and was used to improve both the tail carriage, making it more upright, and to reduce the length of the legs. Seramas do occasionally have Silkie feathers, a trait also known to be carried by the Japanese Bantam.

The resultant offspring from these birds were then interbred, and a number of very tiny birds were hatched during this process. These were the foundation stock of the Serama breed. The size was reduced further, so that today we have birds that can weigh as little as 200gm.

The new breed was named after Sri Rama, the character found in traditional shadow puppet plays and the Indian epic the Ramayana. In these stories Sri Rama was the hero of heroes and goodness incarnate. The breed is amazingly popular in its native country, even inspiring a set of specially issued postage stamps.

Appearance and temperament

Although the true Serama may weigh less than any other breed, it is actually no smaller in body size

Left: Serama male. The bird should have a very short back, a chest that is lifted high and carried well forward, and a tail that should almost touch the back of the head.

than some of the Belgian Barbu de Watermael or d'Anvers (see pages 167-168). The standard for these Belgian breeds has always called for the breed to be as small as possible.

It is often claimed that the breed lacks hardiness. However, it has been reported that when well housed, but without additional heating, Adult Class B Serama, at least, appear able to thrive in a climate where winter temperatures do not drop below -7°C.

The way the Serama holds itself is very important. The chest should be held high and the bird should look bold and full of its own importance. Providing the birds are of good Serama type in the first place, the standard divides the birds into three weight categories:

Class A: Adult males up to 350gm and females up to 325gm.
Class B: Adult males up to 500gm and females up to 425gm.
Class C: Adult males up to 600gm and females up to 525gm.

There is no standard colour for the breed.

The Serama has a very quiet crow and can be kept in limited space, which make it an ideal pet for the urban poultry keeper. Carefully bred, the Serama looks as if it will be a popular new addition to the Bantam scene in the UK.

Points to bear in mind

It is important to emphasise that potential breeders should strive for health as well as size in this breed. It is no use producing very light miniature birds that fail to thrive and grow or are sickly. Fertility of the best Seramas can be poor, with only around 50% of eggs hatching. When over-wintering outdoors, keeping Serama in groups will help them share warmth during very cold weather.

Below: Serama hen. The Serama has a very friendly and tame temperament. This is something judges are required to check for when assessing the breed.

Introduction to Game Breeds

The Game Breeds were all developed for the specific purpose of cockfighting. Just as the racehorse and the greyhound were developed and bred for the amusement of mankind, the natural pugnacity of the Jungle Fowl was refined and honed to increase this natural aggression to the exclusion of all else. Gamefowl are called Hard Feather, as there is no place for soft fluffy feathers here; the birds needed strong, close-fitting feathers that allowed them ease of movement in their task.

Thankfully, cockfighting is becoming less accepted throughout the world, and in 2008 the last two States in the USA that permitted it finally banned the practice. The advent of poultry shows in the nineteenth century made a place for these breeds to remain in existence, to be admired and to compete – not for their lives, but for prizes.

The gamefowl are some of the most remarkable breeds we have. With their sleek lines, close-fitting feathers and bold fiery eyes, the gamefowl display a charismatic spirit found in no other breeds of poultry. These characteristics can still be bred for, without having to pit the birds against each other.

In the UK, cockfighting has been illegal for over 150 years. However, just because something is illegal does not mean it does not still take place. It is a sad fact of life that those

The Modern Game is known for its extreme length of leg.

Old English Game Bantams are a popular show breed.

wishing to preserve and keep the large type of gamefowl for honest exhibition purposes have to maintain exceedingly good security to protect their birds from being stolen by the undesirable types who still practice this so-called sport. Therefore, you will find that breeders of gamefowl are always very wary of strangers who make casual enquiries of them about these fowl.

Let us leave the last word to the writer John Dunton who, as long ago as 1707, wrote: "They then that make themselves SPORT with putting Dogs, Bulls and Cocks to misery, do greatly sin in their pastime. For they make SPORT with exercising cruelty on dumb creatures, which had never been miserable, had not the sins of men made them so".

Asil

It is doubtful that the purest type of this Indian breed was ever exported out of India, as those fowl were too highly prized ever to have been sold to foreigners. However, less valuable specimens of Asil (also spelt Aseel), the most ancient of documented poultry breeds, have made their way to the UK at various times over the last two centuries. And while the UK gamefowl strains may date back a mere few centuries, the Asil has been noted as being in existence for more than a millennium.

In *The Histories of Game Strains*, published in 1928, C. A. Finsterbusch writes "The word 'Asil' denoting (sic) not a mere name to identify a breed but it is an idea summing up nobility, high breeding, courage and desperate gameness."

Appearance and temperament

Strictly a breed for the dedicated expert, the Asil is kept only in small numbers in the UK. There are different types of Asil; the Reza Asil, a small version that weighs less than 3kg, is the most widely kept. This was the type that the gamefowl expert Herbert Atkinson imported into the UK in the early twentieth century.

Even though cockfighting is illegal, the show qualities of the Asil are still judged with the breed's original purpose in mind. Therefore, points are not given for colour; instead, a low tail carriage, pale eye and muscular body shape are desirable when assessing the breed.

The Asil is a very pugnacious, aggressive breed and must be kept separate from other fowl. Asil chicks will start to fight amongst themselves by around eight weeks old and will need separate housing at an early age.

The Asil is widely regarded as the most intelligent of all poultry breeds. The hens are poor layers but good broodies, being very single minded about the task of motherhood.

Points to bear in mind

The breed is not difficult to maintain, but because the Asil is possibly the most sought-after breed for illegal cockfighting, first-class security measures are essential to prevent it being stolen.

Left: *Unlike other gamefowl breeds, Asils should not have rounded shanks but thick square legs, with an indentation down the front of the leg where the scales meet.*

Belgian Game

It was not until the middle of the nineteenth century that this Belgian form of fighting breed became known as Bruges Game. For centuries, large fighting breeds had been developed around the region of Lille along the northern French and Belgian borders. Unlike many other gamefowl, the Belgian Game also became known as a good meat-producing breed. However, although very plentiful, early writers reported that the meat itself was rather "firm and hard" and would need hanging for a while before eating!

Appearance

The Bruges form of Belgian Game is an exceedingly large, very powerful and muscular version of gamefowl, with males often weighing around 5.5kg and females around 4kg. They have a long flat back and very muscular legs. The head has overhanging eyebrows and a strong, stout beak. An infusion of Malay Game is often reported to have been used early on in the creation the breed. The faces of Bruges Game are heavily pigmented and can vary from a mulberry colour to nearly black, which poses

Above: It is classed as a good trait if the hens, here a Bruges type, grow spurs. Egg production is high for a gamefowl breed, at around 150 a year.

the question of where the genes for this particular trait came from.

Various versions of blue plumage have come to dominate the breed. Blue, Blue Birchen, and Blue Yellow Birchen (Lemon Blue) are now the most frequently found colours. Although rare, the breed can be found in both the UK and Europe.

In the latter part of the nineteenth century a different form of Belgian Game was developed in the Liège area by crossing the existing form of Bruges Game out to imported Asian Game, such as Asil or Malay. This Liège form has a longer and more sloping back, as well as longer legs.

Points to bear in mind

As with all large gamefowl breeds, provide extra security measures to prevent the possibility of theft. The breed cannot be recommended for the novice.

Left: The Belgian Game, here a male Bruges type, is strong and powerfully built. It has a very flat back, and the legs should be slate blue in colour.

Indian Game

Developed in Cornwall in the UK, this breed has been widely attributed to Sir Walter Raleigh Gilbert. It seems that in 1846 he stated to a friend that he had formed the breed by crossing imported Asil with Wheaten-coloured Old English Game females. Mention was made of adding Sumatra Game blood at a later date. But this history may well have been one person attempting to lay claim to what was already becoming a popular breed in the West Country, known as the Indian Game.

Although without doubt originally designed for fighting, at which it proved to be a poor performer, the breed soon attracted attention for its ability to add a lot of extra meat when crossed to local fowl.

Every broiler breed today owes its origins to this breed. Once exported to the USA, the name Cornish was used to indicate its place of origin and Cornish crossed with Rock is still one of the most popular meat breeds in the USA today.

Appearance and temperament

The Indian Game is a massive, heavy bird, with thick yellow legs and a very wide breast. The most frequently encountered colour in the UK is the Dark, in which the females have an attractive brown-and-black-laced pattern. In the Jubilee, created in 1897, white replaces the black, resulting in red-and-white lacing. A Bantam form is also available in these colours.

Indian Game are exceedingly poor layers and very eager consumers of large quantities of food. Although the hens want to go broody, they quickly squash their eggs. The hens can also become very bossy towards other hens in the flock. A plus point is that they are not good at getting over even low fences. They are amazing

characters and the hens, especially, can become very tame, particularly if there is any chance of a food treat.

Points to bear in mind

Because of its massive shape and short legs, often compared to those of a bulldog, the Indian Game is very prone to both fertility and hatching problems.

Provide extra-wide pop holes and low perches, as the heaviest birds often suffer from leg and foot problems.

Below: Due to their weight Indian game can sometimes suffer from leg problems. This is a large Dark male.

Ko-Shamo

Of all the Asian Game varieties, this one can be considered as suitable for the careful novice. Although the word 'Ko' means 'small' in Japanese, it is not just a miniaturised version of the larger fighting breed of Shamo (see page 198), but a breed developed in its own right. It first arrived in Europe in 1984 and has been gaining in popularity ever since.

Appearance and temperament

The colour of this fowl is of little importance; it is the shape and style of the bird that matters. A Ko-Shamo should stand very upright and in proportion should consist of one-third head and neck, one-third body and one-third legs. It should have overhanging eyebrows, and a pale-coloured iris to the eye. The head should be well rounded, with a thick, powerful beak.

The Ko-Shamo Is the most amazing character, exceedingly chatty, full of 'attitude' and totally unaware that it stands less than 24cm tall. It has a vast ego for its size. But bear in mind that although it is very sweet-natured towards people, this breed is still a form of Asian Game and should not be mixed with other gamefowl. The wing of a Ko-Shamo

Left: White Ko-Shamo male. The iris colour lightens with age and is referred to as being pearl in colour.

Left: Ko-Shamo Duckwing male. The Ko-Shamo will have bare red skin showing along the breastbone and on the tips of the shoulders.

should always show a split in it. This means that the middle feather, which separates the primary and secondary feathers, is missing. In all other breeds this is classed as a bad fault, but in the Ko-Shamo, and its relation the Chibi, it is regarded as a point of purity.

Points to bear in mind

Although less likely to be stolen than other Asian Game breeds, it is always best to house these birds out of sight of casual observers. A plus point is that they do have a fairly quiet crow and thus are not likely to draw attention to their presence.

Left: Ko-Shamo should have at least four rows of scales around their thick shanks.

Malay

In the Westphalian Landesmuseum in Muenster, Germany, there is a picture by the artist Ludger Tom Ring painted in the year 1570. The painting shows a kitchen table laid out with items awaiting preparation for a wedding feast. Amongst the fish, oranges, crabs and pies, lies the body of what can only be a Wheaten-coloured Malay hen.

The eye of the Malay should have a pale iris and the beak should be strong and powerful.

This is by far the most perfect recording of a poultry breed that has caused a great deal of controversy over the years. The naturalist Coenraad Jacob Temminck proposed that this unusual Asiatic breed must have evolved from an unknown, and now extinct, giant fowl and the idea has given rise to debate ever since.

In the early nineteenth century, the Malay was often referred to as the Great Kulm Fowl, and in 1831 Colonel Sykes brought two cocks and a hen of this breed to the UK. He described the male as standing 26 inches (66cm) tall. This made the Malay a veritable giant amongst the European poultry breeds of the day.

Appearance and temperament

The Malay is often described as a bird of three convex curves, composed of the head and neck, the body, and the tail. The birds' overhanging eyebrows give them a very fierce expression. The eyes of older Malays should have a very pale-coloured iris. This is often not acquired until the birds have been through their first moult. They have a walnut comb and very long, powerful legs.

The Malay is an Asiatic gamefowl breed with a distinctive, hoarse-sounding crow. Birds are best kept in pairs, as they tend to be monogamous. Malay hens are poor layers, often producing only enough eggs to reproduce their kind. The breed is pugnacious by nature and should be kept away from other poultry. Although Malays can become very tame with their keepers, this is certainly not a breed for the novice.

Points to bear in mind

The males can be very protective of their hens and a blow from a Malay's beak will feel akin to being hit with hammer. All Asiatic game will attract the attention of undesirables, who will make determined efforts to steal them for illegal cockfighting.

Left: The Malay is an ancient breed of bird. Bones of Malay-type fowl have been found that date to around 3500 B.C.

Modern Game

Following the ban placed on cockfighting, many people took to pursuing prizes in the newly popular poultry exhibitions instead. As a result of the choices made by judges, birds with longer legs, a higher carriage and tighter-fitting feathers soon started to take the most prizes. This in turn resulted in more people breeding for such points.

These changes went down very badly with those who still believed that all gamefowl should be suitable for fighting; most of the points now being looked for in these fowl rendered the birds totally useless as fighting fowl. Initially, the breed met much opposition but the changes continued. The process was not fast, but by 1870 a breed that looked very different from the original gamefowl was already in existence, going by the name of Exhibition Game. By the 1890s, the fine-boned birds that we know as Modern Game were well established.

Good examples of Modern Game changed hands for hundreds of pounds, a fortune at the time. Yet the glory years of the Large Modern Game were short lived. Fashions change and by 1902 the writer P. Proud reported that the breed was completely out of favour; and out of favour, sadly, it has stayed.

The development of the Bantam form

The Bantam version has been much more successful. The initial development of the breed by John Crossland of Wakefield, Yorkshire was helped on its way to perfection by author and Bantam breeder William Entwistle. Game-type Bantams had been known for a number of years and by the 1870s were starting to grow taller and finer in appearance, more closely resembling the Large Modern Game in appearance.

It is an anomaly that the Modern Game is classed as a Heavy Breed; certainly, the Large Fowl should be very tall and males should weigh around 4kg (but in reality rarely do). However, the Bantams are one of the tiniest and lightest breeds, with full-grown males weighing only around 600gm.

Appearance and temperament

Unlike Old English Game, the colour patterns of Modern Game are very strictly adhered to and the breed is classed as the purest type of exhibition fowl. The Bantams are confident in the extreme; they will be the first to investigate something new, first out of the henhouse door

Left: Large Modern Game Wheaten female. This colour is often found in the Bantams, but seldom seen in Large Fowl.

and often the first to hop into the feed bucket.

The males are far less aggressive amongst themselves than other breeds of gamefowl. However, strains can vary in temperament towards humans. Some strains produce males that are total sweethearts, but others contain males that are akin to Norman Bates from the film Psycho, so choose your birds wisely! Thankfully, the female Bantams are never anything other than friendly, tame, utterly charming and talkative little individuals.

Bantams dislike strong winds, which can blow them off their feet. They can be predated by hawks and cats, so are best kept in sheltered pens. They will often get under your feet, so tread carefully!

Modern Game are an individual taste and, in general, people either love or loathe these tight-feathered, tall, lithe birds. They are not designed as

Right: All Modern Game should display a trait called 'reach', which is an ability to stretch upwards to look very tall. This is a Black-Red male.

Modern Game should have a wedge-shaped body, a short back, very long legs and a small, tightly feathered tail, with no excess of feathering.

Below: *Modern Game Bantam females, with a Partridge (pictured left) and a Birchen. The face of the Birchen should be dark in colour.*

egglayers or meat birds, but do excel at acquiring devotees; novices should be warned that keeping Modern Game Bantams can easily become something of an addiction.

Points to bear in mind

Unfortunately, Large Game breeds always attract undesirable types who will attempt to steal them. Purely for this reason, novices are advised against keeping the Large Fowl version.

Old English Game

The first authentic mention of cockfighting in Britain dates back to the reign of Henry II (1133-1189), but by this time it was obviously a traditional pastime. Cockfighting was restricted and finally banned outright by various acts of Parliament between 1833 and 1845.

When cockfighting was made illegal, the gamefowl that had been bred for centuries for the purpose soon found their way into the newly popular poultry shows as exhibition birds. Within a few years a number of people realised that the birds now being produced for shows were losing the identity, courage and shape of the original fighting fowl and determined that the original form of the breed should be preserved.

In 1882, an exhibition in Cumberland staged a class for the traditional type of fighting game and by 1885 the Old English Game Fowl Club was formed, instigated by the artist and writer Herbert Atkinson. Within two years arguments had broken out regarding the type of the breed required.

Right: *The Oxford Old English Game, here a splendid male, has a lighter and more athletic build than the Carlisle.*

It was Herbert Atkinson who insisted that the breed should be kept strictly to the original type, which later became known as 'Oxfords'.

A second type of Old English Game was also being developed. Over the years these became much broader in shape and developed into a show strain of the breed known as 'Carlisle'. A club for this strain was formed around 1930.

These two strains of Old English Game are still present in large numbers today. Whenever breeders of the two varieties meet there is often very lively discussion and banter about the various merits (or otherwise) of the two types.

In America, where cockfighting remained legal in many States until recently, the pit-game were always kept in the form of a lithe, active, fighting

fowl and many different strains developed. In 1869, J. W. Cooper details no fewer than 73 different strains, all developed with the purpose of fighting in mind. When viewed today, all these American strains bear a similarity to the original Oxford type, as preserved in the UK.

Appearance and temperament

It is important to understand that 'balance' was a vital commodity in the fighting pit and is still an essential quality in the Oxford type. If you watch a judge handle an Oxford Old English Game, he will hold it with the head facing away from him, assessing how well the bird balances when gently held in the hands. When judging a Carlisle version, the judge will hold the bird with the head facing towards him, as it is the overall shape of the bird that matters in this strain.

All gamefowl are found in an amazing array of colours, the names of which defy most logical attempts at explanation. It is said that a good Old English Game can never be a bad colour. This demonstrates that colour is only of very minor importance to the breed. It is the type that is vital. They are not great layers, but many people claim they make superb table birds.

All Large Old English Game are naturally aggressive.

Points to bear in mind

In the UK these birds attract the attention of undesirable types who will make determined efforts to steal them for illegal cockfighting. It is unwise for a novice to attempt to keep any of the large gamefowl.

Below: The Carlisle is a muscular type of fowl, with close-fitting 'hard' feathers. This is a male.

Old English Game Bantams (OEG)

This breed is so popular that it deserves its own entry, separate from the Large Fowl version of the Old English Game. It would be difficult to visit a poultry show anywhere in the UK and not find a selection of these Bantams on show. Confusingly, the breed that became established in type as the Old English Game (usually shortened to OEG) Bantam is a more recent introduction than the breed known as Modern Game. Although Game Bantams had been present at shows for many years, in the latter years of the nineteenth century most Game Bantams were of an exhibition form, and were certainly not looking like miniatures of the original large pit-fighting game. When the first OEG Bantams started appearing at shows around 1898 it was immediately clear that

Rumpless Game

This version of the Old English is lacking the 'parson's nose' and hence has no tail. Although classed as a rare breed, the Rumpless Game is now becoming more popular, frequently taking top awards at shows. It can suffer from mites, so keep a watch for this. It can also suffer from poor fertility. Unlike other OEG, Rumpless Game are not dubbed for showing.

the 'Old English Game Bantam' was a breed of a very different shape and type to the Modern Game.

Appearance and temperament

Writing in 1920, C. A. House described them thus: "A good Old English Game Bantam should be broad in the chest, low on the leg, straight and firm in breast, short in the back, stout in the head and beak, possess a clear, fiery eye of a bold fearless expression". In temperament,

Right: Black-Red OEG Bantam male and Wheaten female. The body shape is strong and powerful.

Rounded legs are essential in this breed. Run your fingers around the legs; they should not feel square.

the OEG Bantam does not contain the same fiery aggressive spark as the Large Fowl version, as it has never been bred for the purpose of cockfighting. The breed is chatty, strong, and hardy. They are great favourites with children, as they are small enough for tiny hands to pick up, and the muscular body is solid enough not to feel either delicate or fragile.

The UK version of the Old English Game is said to resemble the shape of a bullock's heart, and you will often see judges and breeders alike carefully running their hands around the side of a bird to see if it can be moulded into this shape.

The hens make excellent broodies and will happily hatch and rear their own chicks. As layers they can be surprisingly good for a small Bantam that was not developed for the purpose, although of course the eggs suit the size of the bird.

In the UK, the OEG Bantam follows more closely the lines of the Carlisle type of Large Fowl than it does the Oxford, being very wide across the chest, shallow in the keel and with a very tightly whipped tail (meaning the tail feathers are held close together). However, an Oxford version of this breed is now available, with a far more flowing tail, better balance and a more active appearance.

Points to bear in mind

The Old English Game Bantam should prove no more problematic than any other breed of Bantam, which makes it a good choice for the novice. The breed is customarily dubbed (meaning that the comb and wattles are removed) for showing. Consult an experienced game breeder for advice if you wish to show your birds.

OEG Bantam (American type)

In the USA the OEG Bantam has developed along very different lines, carrying a different-shaped body from that found in the UK. Particularly noticeable to British fanciers is the difference in size, as the American versions are up to a third smaller than the British ones.

The OEG in America is a far more stylish-looking breed, adopting its appearance very much from the old pit-game type of fowl. In the Southern States it is customary for the breed to carry rather more tail than is found on birds in the North. Far more emphasis is placed on correct colour patterns, and some very exciting new colours, such as Fawn Duckwing, are being created.

Below: *The tail of American OEG Bantams is large and flowing. This is a Crele male.*

Satsumadori

The Satsumadori only arrived in Europe during the 1980s, but its striking appearance has made it popular with the more specialised exhibition breeders. It is a breed that immediately catches your attention; there is something very powerful, charismatic and flashy about it. It originated in Japan as a fighting breed, produced from a cross between the Long-tailed Shokoku and the gamefowl breed called the Shamo. As such it is an active breed, with long legs and powerful muscles. The tail shape is unusual; unlike other Japanese long-tailed breeds, the tail of the Satsumadori is fanned in appearance.

The most usual colour to find in the UK is the Black, but the breed is also found in Silver Duckwing, Black-Red, Red and White.

Temperament

As befits a breed that was developed for fighting, the Satsumadori still retains a great deal of the original fiery temperament and can be rather aggressive towards its human keepers, as well as other fowl. Keep all Asian game breeds well away from other breeds. Any attempts to integrate them are doomed, invariably resulting in a fatal outcome. The birds are frequently run in pairs of one male and one female, as females can also be aggressive towards one another.

The breed is certainly not kept for its egglaying abilities; in common with most other Asian gamefowl breeds, this can range from exceedingly poor to barely average.

Points to bear in mind

Unfortunately, Satsumadori invariably attract the attention of undesirable types who will make very determined attempts to steal them for illegal cockfighting.

The necks of Asian Game breeds are very strong.

Left: Satsumadori female. This breed is only kept in small numbers at present but is starting to gain in popularity.

Below: Satsumadori male. This breed should have a powerful and muscular build, with close-fitting feathers and a fan-shaped tail.

The shoulders are usually carried well forward and raised, being held slightly away from the body.

Showing, not fighting

While many Hard-feather breeds retain their original pugnacious spirit and athletic shape, selective breeding over many years solely for exhibition points has resulted in dwindling aggression in many strains. Today, breeds such as the tall, slender Modern Game and the short-legged, broad-bodied Indian Game are generally no more aggressive towards other fowl than any other breeds of poultry.

Below: In common with many Asiatic breeds, the Satsumadori has a pea comb. The eye colour varies according to the plumage colour.

Shamo

Although developed as a fighting breed in Japan, where it arrived in the sixteenth century, the origins of this breed lie in Thailand, the name Shamo being a derivation of the Japanese word for that country. The breed is officially known in Japan as the O-Shamo and is the largest of the Shamo varieties found there.

Appearance and temperament

The Shamo is a tall, gaunt, fierce-looking breed with a pea comb, prominent shoulders and a sloping back. It is noticeable that the Asian game breeds lack feathers at both the keel (breastbone area) and also at the points of the shoulders. Both these

areas will have bare red skin showing, with no trace of feathers there.

The beak is very strong and broad, and the earlobes and wattles should be either very small or absent. The neck is long and the body very muscular. The breed is exceedingly pugnacious, retaining every ounce of its fighting instincts and abilities. It is not a breed to be undertaken unless you have plenty of poultry-keeping experience.

Points to bear in mind

As with all Asian Game, the probability of theft is remarkably high.

Below: Shamo male. When viewed in profile the slope of the back becomes very pronounced.

Note the bare red skin, both in the breastbone area and at the points of the shoulders.

Left: Shamo female. The females do not readily accept new hens and the breed is often kept in pairs.

The tail will follow the downward slope of the back.

Long-Crowing Breeds

Whereas no-one has ever gone out of their way to develop a silent cockerel, all around the world people have made every effort to produce breeds that are very vocal. Most of these breeds look just like any other type of poultry, the difference only becoming obvious when they open their beaks. Long-Crowing breeds include the Bergische Kraeher from Germany, the Berat from Albania, a bass crower, and the Denizli, developed in Turkey, that 'sings' for anything up to 35 seconds.

The Totenko, the Tomaru and the Kurokashiwa all hail from Japan, but perhaps the strangest of all these breeds is the Koeyoshi, also from Japan. It is a huge, gaunt-looking bird that resembles the largest of the Asian gamefowl. It can produce a deep bass crow for around 20 seconds without drawing breath and with its beak mostly closed!

In their native countries crowing competitions take place, where the birds are placed on a table

Above: *The Kurokashiwa from Japan has both a long crow and a long tail.*

The Yamato Gunkei

Other Asian Game Breeds include the Yamato Gunkei and the Chibi. If ever a breed could be said to resemble a dinosaur, then it is the Yamato Gunkei. It is very broad and its face is covered in wrinkles and flesh. It weighs around 2kg. The older the bird, the more wrinkles it acquires. It is reputed to dislike cold weather. This very unusual and rare breed was first imported into Europe in 1984, but as with all Asian Game, it is not a breed for the novice.

The Chibi is a dwarf version of the Yamato Gunkei, weighing only around 1kg. It is not known if any Chibi are currently available in the UK, although they do exist in Europe.

Left: *The wrinkled faces on the Yamato Gunkei become more pronounced with age.*

and the sound of the crow is judged for tone, length and evenness. As yet, no crowing competitions have been held in the UK.

Although some of these breeds are now found in the UK they are still understandably scarce, so acquiring stock will be a problem. Unless you have very tolerant or deaf neighbours, the long-crowers are best left in the hands of people who live deep in the countryside. You could search out websites with links to recordings of the breeds – they have to be heard to be believed!

199

Glossary

Balance A term used to describe a gamefowl with a build that would prove agile and stable when moving quickly. Used when describing breeds such as Oxford Old English Game.

Barred A plumage pattern of black-and-white barring across the feathers.

Beard A group of feathers found under the beak (*see* the Houdan).

Birchen Generally refers to a two-toned colour pattern (usually black-and-white), where the breast is laced. Best seen in Birchen Modern Game.

Black-Red *see* Partridge

Blue A pigeon blue-grey colour. In reality, a dilute form of black (*see* Andalusian).

Boule A neck hackle that forms an unusually thick manelike effect of feathers, best seen in breeds such as the Orloff.

Clean legs Legs without feathers on them.

Columbian A colour pattern where the neck hackle carries a darker central stripe, usually black, surrounded by a white edge to the feather. The tail and wings will also carry the darker colour. This is best seen in breeds such as the Light Sussex or Light Brahma. Buff or other colours can replace the white, depending on the colour variety.

Cuckoo A type of barred plumage pattern in which the barring definition is softer and indistinct.

Duckwing A plumage colour where the red-orange and brown colour of the partridge pattern is replaced by white, or in the case of Gold Duckwing, where the red is replaced by a light straw colour. In addition, a Duckwing male will have a steel blue colour in the wing bars.

Duplex, or horn, comb A comb that resembles horns and looks rather similar to the letter V. Usually found in breeds with crests.

Hard Feather A term used to describe the game breeds that were originally used in cockfighting. The feathers are tight and sleek, with no fluff, and are held very close to the body.

Hen feathered A male bird that fails to develop the usual curved sickle feathers in the tail and the larger hackle feathers normally found in the cock bird. Instead, a hen-feathered cock will grow shorter feathers, similar to those found in the hen.

Lacing A distinct line of different colour found around the edge of a feather.

Lavender A true breeding form of the blue colour, which reproduces true to colour.

Leader The pointed end that is found at the back of a rose comb.

Leaf comb A very unusual comb that looks like a leaf (*see* the Houdan for an example).

Mottled A plumage pattern that has a white spot at the end of each feather.

Muffling Feathers found around and alongside the beak, giving a whiskery appearance.

Partridge The natural wild pattern found in jungle fowl, where the male has a black breast and red or orange neck hackle and the female has a light brown body with a salmon breast. Males are usually referred to as Black-Reds.

Pea comb (or triple comb) Three small single combs joined together at the base, the middle of the three being slightly higher than the outside ones (*see* the Brahma for an example).

Pencilled Confusingly, there are two types of plumage that at different times were given the same name. It can either refer to small stripes straight across the feather in bands, such as found in the Friesian, or the term can refer to multiple concentric lines that follow the outline of the feather, such as are found in the Dark Brahma female. Its use depends on the breed it is describing

Pile (Pyle) A reversal of the partridge pattern, with white replacing the black on the male and the brown on the female.

Reach The ability of a bird to stretch itself upwards and stand very tall – a vital attribute in breeds such as Modern Game.

Rose comb A broad comb on which the surface is covered with many small, fleshy bumps, the back part ending in a spike known as a leader.

Side sprig Spikes that grow outwards from the side of single combs – a fault in most breeds.

Single comb The upright comb found in breeds such as Minorca, Orpington or Welsummer.

Spangling A spot of colour found on the end of a feather, which is a different colour from the rest of the feather.

Split wing An unwanted gap between the primary and secondary feathers. A fault in all breeds except a few of the small Asian Game breeds.

Soft Feather Refers to all breeds of poultry that were not created for the purpose of cockfighting.

Undercolour/Underfluff The colour found on the base and the lower part of the feather that does not normally show. It is only visible when the feathers are parted.

Vulture hocks Stiff feathers that grow outwards and downwards from the hock joint.

Wheaten The males are similar to the Partridge plumage pattern but the body colour of the females is replaced with a soft creamy fawn, supposed to be the colour of new wheat.

Wing bars A line of differently coloured feathers, often a steely blue colour, that appears across the middle of the wing in certain plumage patterns such as Duckwing.

Workings The small bumps found on top of a rose comb.

General index

Page numbers in **bold** indicate major entries, including photos; *italics* refer to captions, annotations and panels; plain type indicates other text entries.

Index to Chicken Breeds

Bibliography

An Encyclopædia of Agriculture 1826, John Claudius Loudon (Longman, Rees, Orme, Brown, and Green)

Araucana Poulterers Handbook, David Caudill (The International Collonca Society, 1975)

Ameraucana Breeders Club: www.ameraucana.org

Association for Promotion of Belgian Poultry Breeds: http://users.telenet.be/jaak.rousseau/english%20version/CLUB/CLUB.HTM

Athenian Sport, John Dunton (B. Bragg, 1707)

Avicultura website: www.aviculture-europe.nl/indexUK.html

Bantams and How to Keep Them, C. A. House (Read Books, Reprint 2005)

British Poultry Standards 5th Edition (Blackwell Science, 1997)

Bulletin of the Texas Ornithological Society Vol. 39, No. 1, January 2006, Pages 1-32

Chattering On Gallus William John Plant, 1997: www.summagallicana.it/chatteringongallus/index.htm

Chickens for Use and Beauty, H. S. Babcock, 1890 pp. 47-60

The Century; a popular quarterly. Vol 40, Issue 1

Country Smallholding, September 2000-Present

Empty Shells, Thea Snyder Lowry (Manifold Press, 2000)

Fancy Fowl Magazine, October 1981-Present

Fowls; A Treatise on the Principle Breeds, John Baily (Henningham and Hollis, 1860)

Game Fowls, Their Origin and History, J. W. Cooper, 1869

General Report of the Agricultural State, And Political Circumstances, Of Scotland, 1814

Handbook of Avian Hybrids of the World, Eugene M. McCarthy (Oxford University Press)

How to Keep Hens for Profit, C. S. Valentine (Macmillan Co. New York, 1910)

Miners Domestic Poultry Book T. B. Miner (Geo. W. Fisher, 1853)

Moubray's Treatise on Domestic and Ornamental Poultry 1837, John Lawrence (Joseph Breck and Co.)

National Geographic, December 1970

Ornamental, Aquatic, and Domestic Fowl, And Game Birds James Joseph Nolan, 1850

Outdoor Sports and Games, Claude Harris Miller (Plain Label Books, 1911)

Poultry Breeding and Genetics, Roy D. Crawford (Elsevier, 1990)

Poultry Breeding and Production, Edward Brown (Ernest Benn Limited, 1929)

Poultry Breeds and Management, David Scrivener (The Crowood Press Ltd., 2008)

Poultry Club Of Great Britain Year Books 1977-2007

Poultry Tribune, 1941

Practical Poultry Magazine, March 2004-Present

Rare Poultry Breeds, David Scrivener (The Crowood Press Ltd., 2006)

Rare Poultry Society Newsletters

The American Breeds of Poultry, Frank L. Platt (Marcel Press, Reprint 2008)

The Book of Poultry, Lewis Wright (Cassell and Company, 1885)

The Complete Encyclopaedia of Chickens, Ester Verhoef and Aad Rijs (Rebo International, 2006)

The Cultivator 1857, New York State Agricultural Society

The Field Magazine: The Fighting Cock and Other Fowl, Rex Woods, January 1965

The History of Hen Fever, Geo. P. Burnham (James French and Company, 1855)

The Perfected Poultry of America, T. F. McGraw (The Howard Publishing Co., 1907)

The Poultry Yard, Elizabeth Watts (Routledge & Sons, London, 1870)

The Soil Association: www.soilassociation.org

The Te Aroha News, November 1889

The UK Serama Club: www.ukseramaclub.co.uk

World Poultry Congress: Report of Proceedings, Ottawa, Canada, 1927

Credits

Unless otherwise stated, photographs have been taken by Geoff Rogers © Interpet Publishing Ltd.

The publishers would like to thank the following for providing images, credited here by page number and position: (B) Bottom, (T) Top, (C) Centre (BL) Bottom left, etc.

Frances Bassom: 9(TL), 71, 72, 79(TL, TR), 83, 105(Both), 137, 138, 161(TR), 178(BL), 179(Both), 182, 183, 185

Matthew Burrard-Lucas: 8

Forsham Cottage Arks: 22(L), 23(TL)

Yvonne Hillsden: 129(Both)

Alison Ingram, Warnham Welsummers: 118(B)

Julia Keeling: 199(Both)

Johan Opsomer, Belgium: 86(TR), 87(Both), 90, 92, 93, 94, 169(TR), 186(Both)

Rachel Pitt: 2TR (Contents), 39(TR), 51(BR)

Hans L. Schippers, Holland: 84, 91, 97, 98(Both) 99, 101(BL), 113, 125, 131, 145(TR), 159(BR), 168(BL)

Neil Sutherland © Interpet Publishing Ltd: 17(Both), 22(R), 28(T, Both), 30(TR, Both), 33(TR), 36(TR), 37(T), 46, 57(BL), 69(B), 75(BR), 107(Both), 149(TR)

John Tarren (David Scrivener Archive): 100, 101(TR), 104, 106, 108, 109, 110, 111, 130(Both), 145(BL), 146, 192, 195

Author's acknowledgements

I would like to thank everyone who has been kind enough to assist with the creation of this book. In particular, I would like to thank Andy Tate for the time and assistance he has me given in preparing the book; Dave Scrivener for suggesting I write the book in the first place; and Priscilla Middleton for all the advice given freely over the years. Finally, but very importantly, my mother the late Beryl Johnston, who always encouraged an enquiring mind; and who firmly informed me that *proper chickens* were called Buff Orpingtons.

Publishers acknowledgements

The publishers would like to thank the following for their help during the preparation of this book:

Tony, Sue and Kate Beardsmore, Cathy Burton, Judith Burton, Colchester Poultry Club, James Firth, Nicola Firth, Stephen Flory at The Henhouse Garden Co., David Francis, Lester Frenzel and Daniel, Tim Fuller, Lana Gazder, Simon and Tracey Hayter, Charlotte Heales, Sally Hutton, LLoyd Ince, Mayfield Poultry, Priscilla Middleton, Rachel and Andrew Pitt, David Scrivener, Jill Tait, Joy and Steve Thorpe, Mulberry Mill Stud, Suffolk.

Publisher's note

The information and recommendations in this book are given without guarantee on the part of the author and publishers, who disclaim any liability with the use of this material.